Henry Lee

The Vegetable Lamb of Tartary

A Curious Fable of the Cotton Plant

Henry Lee

The Vegetable Lamb of Tartary
A Curious Fable of the Cotton Plant

ISBN/EAN: 9783744674324

Printed in Europe, USA, Canada, Australia, Japan

Cover: Foto ©berggeist007 / pixelio.de

More available books at **www.hansebooks.com**

THE "BAROMETZ," OR "TARTARIAN LAMB."

After Joannes Zahn.

THE VEGETABLE LAMB

OF

TARTARY;

A Curious Fable of the Cotton Plant.

TO WHICH IS ADDED

A SKETCH OF THE HISTORY OF COTTON AND
THE COTTON TRADE.

BY

HENRY LEE, F.L.S., F.G.S., F.Z.S.,

SOMETIME NATURALIST OF THE BRIGHTON AQUARIUM,
AND
AUTHOR OF 'THE OCTOPUS, OR THE DEVIL-FISH OF FICTION AND OF FACT,'
'SEA MONSTERS UNMASKED,' 'SEA FABLES EXPLAINED,' ETC.

ILLUSTRATED.

LONDON:
SAMPSON LOW, MARSTON, SEARLE, & RIVINGTON,
CROWN BUILDINGS, 188, FLEET STREET.
1887.

All Rights reserved.

CONTENTS.

CHAPTER I.
THE FABLE AND ITS INTERPRETATION 1

CHAPTER II.
THE HISTORY OF COTTON AND ITS INTRODUCTION INTO EUROPE . 63

APPENDIX 97

LIST OF ILLUSTRATIONS.

FIG.   PAGE

The "Barometz," or "Tartarian Lamb."—*After Joannes Zahn* . . . . *Frontispiece*

1.—The Vegetable Lamb Plant.—*After Sir John Mandeville* . 3

2.—Portrait of the "Barometz," or "Scythian Lamb."—*After Claude Duret* 9

3.—Adam and Eve admiring the Plants in the Garden of Eden. The "Vegetable Lamb" in the background.—*Fac-simile of the Frontispiece of Parkinson's "Paradisus"* . 19

4.—Rhizome of a Fern, shaped by the Chinese to represent a tan-coloured Dog, and laid before the Royal Society by Sir Hans Sloane as a Specimen of the "Barometz," or "Tartarian Lamb."—*From the 'Philosophical Transactions,' vol. xx., p.* 861 25

5.—Rough model of a tan-coloured Dog, shaped by the Chinese from the Rhizome of a Fern, and submitted to the Royal Society by Dr. Breyn as a Specimen of the "Scythian Vegetable Lamb," or Borametz.—*From the 'Philosophical Transactions,' No.* 390 31

6.—The "Borametz," or "Scythian Lamb."—*From De la Croix's 'Connubia Florum'* 37

7.—A Cotton-pod 61

PREFACE.

THE fable of the existence of a mysterious "plant-animal" variously entitled "*The Vegetable Lamb of Tartary*," "*The Scythian Lamb*," and "*The Barometz*," or "*Borametz*," is one of the curious myths of the Middle Ages with which I have been long acquainted. Until the year 1883, not having given serious thought to it, or made it a subject of critical examination, I had been content to accept as correct the explanation of it now universally adopted; namely, that it originated from certain little lamb-like toy figures ingeniously constructed by the Chinese from the rhizome and frond-stems of a tree-fern, which, from its identification with the object of the fable, has received the name of *Dicksonia Barometz*. But during my researches in the works of ancient writers when preparing the manuscript of my two books, '*Sea Monsters Unmasked*,' and '*Sea Fables Explained*,' I came upon passages of old authors which convinced me that these toy "lambs" made from ferns by the Chinese had no more connexion with the story of "*The Vegetable Lamb*" than the artificial mermaids so cleverly constructed by the Japanese were the cause and origin of the ancient and world-wide belief in mermaids. Subsequent investigations have confirmed this opinion.

I have found that all of these old myths which I have been able to trace to their source have originated in a perfectly true statement of some curious and interesting fact; which statement has been so garbled and distorted, so misrepresented and perverted in repetition by numerous writers, that in the course of centuries its original meaning has been lost, and a monstrous fiction has been substituted for it. "Truth lies at the bottom of a well," says the adage; and in searching for the origin of these old myths and legends, the deeper we can dive down into the past the greater is the probability of our discovering the truth concerning them. To obtain a clue to the identity of "*The Scythian Lamb*" we must consult the pages of historians and philosophers who lived and wrote from eighteen to sixteen centuries before Sir John Mandeville published his version of the story; and, having there found set before us the real "*Vegetable Lamb*" in all its truthful simplicity and beauty, we shall be able to recognise its form and features under the various disguises it was made to assume by the wonder-mongers of the Middle Ages.

I venture to believe that the reader who will kindly follow my argument (p. 42, *et seq.*) will agree with me that the rumour which spread from Western Asia all over Europe, and was a subject of discussion by learned men during many centuries, of the existence of "a tree bearing fruit, or seed-pods, which when they ripened and burst open were seen to contain little lambs, of whose soft white fleeces Eastern people wove material for their clothing," was a plant of far higher importance to mankind than the paltry toy animals made by the Chinese from the root of a fern, of which gew-gaws only four specimens are known to have been brought to this country. It seems to me clear and indisputable that the rumour referred to the cotton-pod, and originated in the first introduction of cotton and the fabrics woven from it into Eastern Europe.

It will be seen that the explanation of the process by

which the truthful report of a remarkable fact was in time perverted into the detailed history of an absurd fiction is very easy and intelligible.

As this little book was originally intended for publication, like its predecessors before-mentioned, as a hand-book in connection with the Literary Department of the South Kensington Exhibitions, I have treated in a separate chapter of the history of cotton, its use by ancient races in Asia, Africa, and America, and its gradual introduction amongst the nations of Europe. The various stages of its progress Westward were so distinctly and intimately dependent on many remarkable events in the world's history, by which its advance was alternately retarded and facilitated, that the annals of the "*vegetable wool*" which holds so important a place amongst the manufacturing industries of Great Britain are hardly less romantic than the fable of "*The Vegetable Lamb*," which was its forerunner.

HENRY LEE.

SAVAGE CLUB.
May, 1887.

THE
VEGETABLE LAMB OF TARTARY

A CURIOUS FABLE OF THE COTTON PLANT.

CHAPTER I.

THE FABLE AND ITS INTERPRETATION.

AMONGST the curious myths of the Middle Ages none were more extravagant and persistent than that of the "Vegetable Lamb of Tartary," known also as the "Scythian Lamb," and the "Borametz," or "Barometz," the latter title being derived from a Tartar word signifying "a lamb." This "lamb" was described as being at the same time both a true animal and a living plant. According to some writers this composite "plant-animal" was the fruit of a tree which sprang from a seed like that of a melon, or gourd; and when the fruit or seed-pod of this tree was fully ripe it burst open and disclosed to view within it a little lamb, perfect in form, and in every way resembling an ordinary lamb naturally born. This remarkable tree was supposed to grow in the territory of "the Tartars of the East," formerly called "Scythia"; and it was said that from the fleeces of these "tree-lambs," which were of surpassing whiteness, the natives of the country where they were found wove materials for their garments and "head-dress." In the course of time another version of the story was circulated, in which the lamb was not described as being the fruit of a tree, but as being a living lamb attached by its navel to a short stem rooted in the earth. The stem, or stalk, on which the lamb was thus suspended above the ground was sufficiently flexible to allow the animal to bend downward, and browze on the herbage within its reach. When all the grass within the length of its tether had been consumed the stem withered and the lamb died. This plant-lamb was reported to have bones, blood, and delicate flesh, and to be a favourite food of wolves, though no other carnivorous animal would attack it. Many other details were given concerning it, which will be found mentioned in the following pages. This legend met with almost universal credence from the thirteenth to the seventeenth centuries, and, even then, only gave place to an explanation of it as absurd and delusive as itself. Following the outline sketched in the preface, I shall, in this chapter, lay before the reader the story of the "Barometz" or "Vegetable Lamb," as related by various writers, and shall then give my reasons for assigning to the fable an interpretation very different from that which has been hitherto accepted as the true one.

The story of a wonderful plant which bore living lambs for its fruit, and grew in Tartary, seems to have been first brought into public notice in England in the reign of Edward III., by Sir John Mandeville, the "Knyght of Ingelond that was y bore in the toun of Seynt Albans, and travelide aboute in the worlde in many diverse countreis, to se mervailes and customes of countreis, and diversiteis of

FIG. 1.—THE VEGETABLE LAMB PLANT.

After Sir John Mandeville.

This plate illustrates that version of the Fable by which the "Vegetable Lamb" is represented as contained within a fruit, or seed-pod, which, when ripe, bursts open, and discloses the little lamb within it.

folkys, and diverse shap of men and of beistis." In the 26th chapter of the book in which he "wrot and telleth all the mervaile that he say," and which he dedicated to the King, he treats of "the Countreis and Yles that ben beʒond the Lond of Cathay, and of the Frutes there"; and amongst the curiosities he met with in the dominions of the "Cham" of Tartary he mentions the following:—

"Now schalle I seye ʒou semyngly of Countrees and Yles that ben beʒonde the Countrees that I have spoken of. Wherefore I seye you in passynge be the Lond of Cathaye toward the high Ynde, and towards Bacharye, men passen be a Kyngdom that men clepen Caldilhe: that is a fair Contree. And there growethe a maner of Fruyt, as though it weren Gowrdes: and whan thei ben rype men kutten hem ato, and men fynden with inne a lytylle Best, in Flesche, in Bon and Blode, as though it were a lytylle Lomb with outen Wolle. And Men eten both the Frut and the Best; and that is a great Marveylle. Of that Frute I have eaten; alle thoughe it were wondirfulle, but that I knowe wel that God is marveyllous in his Werkes."*

Sir John Mandeville appears to have never previously heard of this strange plant, but reports of its existence under various phases may be traced back, as we shall presently see, to a date at least eighteen hundred years earlier than that of his mention of it. As it is in the works of these older writers that we shall find the long-sought key of the mystery, we will set them aside for the present and follow the growth and dissemination of the fable.

Claude Duret, of Moulins, who, in his 'Histoire Admirable des Plantes (1605),' devotes to it a chapter entitled "The Boramets of Scythia, or Tartary, true

* 'The Voiage and Travaile of Sir John Maundevile, Knt.' See Appendix A.

Zoophytes or plant-animals; that is to say, plants living and sensitive like animals," therein says:—

"I remember to have read some time ago in a very ancient Hebrew book entitled in Latin the *Talmud Ierosolimitanum*, and written by a Jewish Rabbi Jochanan, assisted by others, in the year of salvation 436, that a certain personage named Moses Chusensis (he being a native of Ethiopia) affirmed, on the authority of Rabbi Simeon, that there was a certain country of the earth which bore a zoophyte, or plant-animal, called in the Hebrew '*Jeduah.*' It was in form like a lamb, and from its navel grew a stem or root by which this zoophyte or plant-animal was fixed, attached, like a gourd, to the soil below the surface of the ground, and, according to the length of its stem or root, it devoured all the herbage which it was able to reach within the circle of its tether. The hunters who went in search of this creature were unable to capture or remove it until they had succeeded in cutting the stem by well-aimed arrows or darts, when the animal immediately fell prostrate to the earth and died. Its bones being placed with certain ceremonies and incantations in the mouth of one desiring to foretell the future, he was instantly seized with a spirit of divination, and endowed with the gift of prophecy."

As I was unable to find in the Latin translation of the Talmud of Jerusalem the passage mentioned by Claude Duret, and was anxious to ascertain whether any reference to this curious legend existed in the Talmudical books, I sought the assistance of learned members of the Jewish community, and, amongst them, of the Rev. Dr. Hermann Adler, Chief Rabbi Delegate of the United Congregations of the British Empire. He most kindly interested himself in the matter, and wrote to me as follows:—

"It affords me much gratification to give you the information you desire on the Borametz. In the Mishna *Kilaim*, chap. viii. § 5 (a portion of the Talmud), the passage occurs:—'Creatures called *Adne Hasadeh* (literally, "lords of the field") are regarded as beasts.' There is a variant reading,—*Abne Hasadeh* (stones of the field). A commentator, Rabbi Simeon, of Sens (died about 1235), writes as follows on this passage:—'It is stated in the Jerusalem Talmud that this is a human being of the mountains: it lives by means of its navel: if its navel be cut it cannot live. I have heard in the name of Rabbi Meir, the son of Kallonymos of Speyer, that this is the animal called '*Jeduah.*' This is the '*Jedoui*' mentioned in Scripture (lit. *wizard*, Leviticus xix. 31); with its bones witchcraft is practised. A kind of large stem issues from a root in the earth on which this animal, called '*Jadua,*' grows, just as gourds and melons. Only the '*Jadua*' has, in all respects, a human shape, in face, body, hands, and feet. By its navel it is joined to the stem that issues from the root. No creature can approach within the tether of the stem, for it seizes and kills them. Within the tether of the stem it devours the herbage all around. When they want to capture it no man dares approach it, but they tear at the stem until it is ruptured, whereupon the animal dies.' Another commentator, Rabbi Obadja of Berbinoro, gives the same explanation, only substituting—'They aim arrows at the stem until it is ruptured,' &c. The author of an ancient Hebrew work, Maase Tobia (Venice, 1705), gives an interesting description of this animal. In Part IV. c. 10, page 786, he mentions the Borametz found in Great Tartary. He repeats the description of Rabbi Simeon, and adds what he has found in 'A New Work on Geography,' namely, that 'the Africans (*sic*) in Great Tartary, in the province of Sambulala, are enriched by means of seeds like the seeds of gourds, only shorter in size, which grow and blossom like a stem to the navel of an animal which is called *Borametz* in their language, i.e. '*lamb*,' on account of its resembling a lamb in all its limbs, from head to foot; its hoofs are cloven, its skin is soft, its wool is adapted for clothing, but it has no horns, only the hairs of its head, which grow, and are intertwined like horns. Its height is half a cubit and more. According to those who speak of this wondrous thing, its taste is like the flesh of fish, its blood as sweet as honey, and it lives as long as there is herbage within reach of the stem, from which it derives its life. If the herbage is destroyed or perishes, the animal also dies away. It has rest from all beasts and birds of prey, except the wolf, which seeks to destroy it.' The author concludes by expressing his belief, that this account of the animal having the shape of a lamb is more likely to be true than that it is of human form."

We have an interesting record of another journey into Tartary, undertaken almost simultaneously with that of Sir John Mandeville, by Odoricus of Friuli, a Minorite friar belonging to the monastery of Utina, near Padua. The exact date of his departure on his travels is not mentioned, but he returned home in 1330, and the history of his adventures and observations* was written in the month of May of that year—thus taking precedence by about thirty years of the narrative of the old English traveller.

Odoricus, describing his visit to the country of the "Grand Can," says:—"I heard of another wonder from persons worthy of credit; namely, that in a province of the

---

* 'The Journall of Frier Odoricus of Friuli, one of the order of the Minorites, concerning strange things which he saw amongst the Tartars of the East.'—' Hakluyt Collection of Early Voyages,' vol. ii. 1809. See Appendix B.

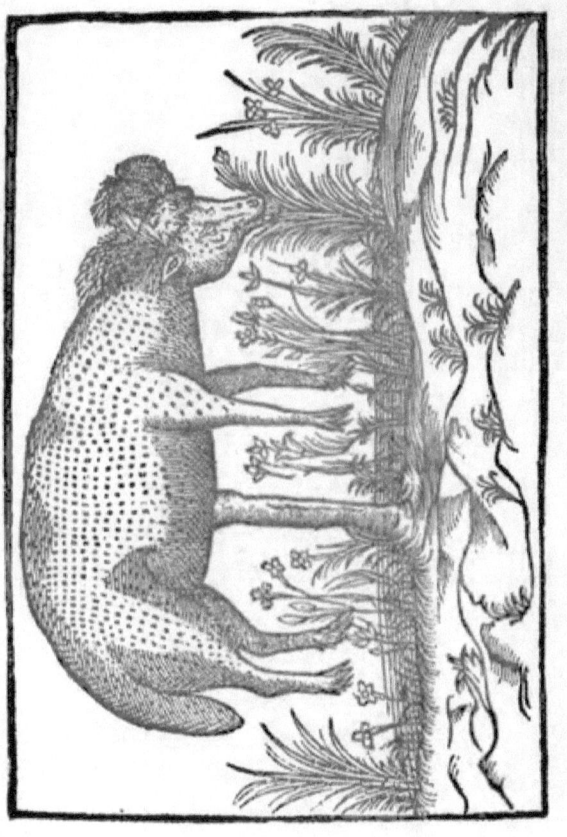

FIG. 2.—PORTRAIT OF THE "BAROMETZ," OR "SCYTHIAN LAMB."

After Claude Duret.

said Can, in which is the mountain of Capsius* (the province is called 'Kalor'), there grow gourds, which, when they are ripe, open, and within them is found a little beast like unto a young lamb, even as I myself have heard reported that there stand certain trees upon the shore of the Irish Sea bearing fruits like unto a gourd, which at a certain time of the year do fall into the water and become birds called Bernacles; and this is true."

In the sixteenth and seventeenth centuries the "Scythian Lamb" was made a subject of investigation and argument by some of the most celebrated writers of that period.

Fortunio Liceti, Professor of Philosophy at Padua, writing in 1518,† gives his complete credence to the story of the little beast like a lamb found within a fruit-pod when it bursts from over-ripeness; and besides the above passage from Odoricus quotes another, by which it would appear that the worthy friar afterwards himself saw this botanical curiosity, and described it as being "as white as snow." I have been unable to find this paragraph in the Hakluyt edition of Odoricus's travels.

Juan Eusebio Nieremberg, however, in his '*Historia Naturæ*' (Antwerp, 1605), also quotes these two passages, and in exactly the same words. He probably copied them from Liceti, and not from the original.

Sigismund, Baron von Herberstein, who, in 1517 and 1526, was the ambassador of the Emperors Maximilian I. and Charles V. to the "Grand Czard, or Duke of Muscovy," in his 'Notes on Russia,'‡ gives further details of this

---

* Probably an error of transcription for "Caspius." The mountain of Caspius (now Kasbin) is about eighty miles due south of the Caspian Sea, and in Persian territory, near Teheran.

† '*De Spontaneo Viventium Ortu*,' lib. 3, cap. 45.

‡ '*Rerum Muscoviticarum Commentarii*,' 1549. See Appendix C.

"vegetable-animal." He writes:—" In the neighbourhood of the Caspian Sea, between the rivers Volga and Jaick, formerly dwelt the kings of the Zavolha, certain Tartars, in whose country is found a wonderful and almost incredible curiosity, of which Demetrius Danielovich, a person in high authority, gave me the following account; namely, that his father, who was once sent on an embassy by the Duke of Muscovy to the Tartar king of the country referred to, whilst he was there, saw and remarked, amongst other things, a certain seed like that of a melon, but rather rounder and longer, from which, when it was set in the earth, grew a plant resembling a lamb, and attaining to a height of about two and a half feet, and which was called in the language of the country 'Borametz,' or 'the little Lamb.' It had a head, eyes, ears, and all other parts of the body, as a newly-born lamb. He also stated that it had an exceedingly soft wool, which was frequently used for the manufacturing of head-coverings. Many persons also affirmed to me that they had seen this wool. Further, he told me that this plant, if plant it should be called, had blood, but not true flesh: that, in place of flesh, it had a substance similar to the flesh of the crab, and that its hoofs were not horny, like those of a lamb, but of hairs brought together into the form of the divided hoof of a living lamb. It was rooted by the navel in the middle of the belly, and devoured the surrounding herbage and grass, and lived as long as that lasted; but when there was no more within its reach the stem withered, and the lamb died. It was of so excellent a flavour that it was the favourite food of wolves and other rapacious animals. For myself," adds the Baron, "although I had previously regarded these Borametz as fabulous, the accounts of it were confirmed to me by so many persons worthy of credence that I have thought it right to describe it; and this with the less hesitation because I was told by Guillaume Postel,* a man of much learning, that a person named Michel, interpreter of the Turkish and Arabic languages to the Republic of Venice, assured him that he had seen brought to Chalibontis (now Karaboghaz), on the south-eastern shore of the Caspian Sea, from Samarcand and other districts lying towards the south, the very soft and delicate wool of a certain plant used by the Mussulmans as padding for the small caps which they wear on their shaven heads, and also as a protection for their chests. He said, however, that he had not seen the plant, nor knew its name, except that it was called 'Smarcandeos,' and was a zoophyte, or plant-animal. The numerous descriptions given to him," he added, "differed so little that he was induced to believe that there was more truthfulness in this matter than he had supposed, and to accept it as a fact redounding to the glory of the Sovereign Creator, to whom all things are possible."

Shortly after the publication of the above narrative by Sigismund von Herberstein, and probably in allusion to it, Girolamo Cardano, of Pavia, carefully discussed the phenomenon in question in his work 'De Rerum Naturâ,'† printed at Nürnberg in 1557. He endeavoured to expose the absurdity of the statements made concerning this "animal-plant," and explained the physical impossibility of its existence in the manner described. He argued that if it had blood it must have a heart, and that the soil in which a plant grows is not fitted to supply a heart with movement and vital heat. He also pointed out that embryo animals, especially, require warmth for their development from the

* Author of 'Liber de Causis, seu de Principiis et Originibus Naturæ,' &c.
† Lib. vi. cap. 22.

ovum, which they could not obtain if raised from a seed planted in the earth, demonstrating clearly enough that no warm-blooded animal could exist thus organically fastened to the earth. In reply, however, to a possible question suggested by himself, why there should be no plant-animal on land, seeing that there are zoophytes in the sea, he, with the weakness and indecision which were innate in his character, admitted that "where the atmosphere was thick and dense there might, perhaps, be a plant having sensation, and also imperfect flesh, such as that of mollusks and fishes."

This weak point in his argument laid him open to the criticism of his relentless enemy, Julius Cæsar Scaliger. Always on the watch to wound and harass Cardano with cutting satire and irritating gibes, this caustic persecutor lost no time in making his attack. In one of his "*Exercitationes*"[*] he thus personally addressed the object of his sneering disparagement :—

"You may regard as beyond ridicule this wonderful Tartar plant. The most renowned of the Tartar hordes of the present day, by its reputation, its antiquity, and its nobility, is that of the Zavolha. These people sow a seed like that of the melon, but rather smaller, from which springs and grows out of the earth a plant which they call 'Borametz,' *i.e.* 'the Lamb.' This plant grows to the height of three feet in the likeness of a real lamb, having feet, hoofs, ears, and a head perfect with the exception of horns, instead of which the plant has hairs in the form of horns. Its skin is soft and delicate, and is used in Tartary for head-gear. The internal pulp is said to be like the flesh of the cray-fish, and to have an agreeable flavour ;

---

[*] '*Exotericarum Exercitationum*,' lib. xv., "*De Subtilitate*"; *ad Hieronymum Cardanum Exercit.* 181, cap. 29. Frankfort, 1557. See Appendix D.

but if an incision be made, real blood flows from it. The root or stalk which rises from the earth is attached to the navel of the lamb, and (which is more remarkable) whilst the plant is surrounded with herbage it lives as does a lamb, but as soon as it has consumed all within its reach it withers and dies. This does not happen by the arrival of the plant at any definite period of its growth, for it has been found by experiment that if the grass around it be removed it perishes. Another most curious circumstance connected with it is that wolves will eat it with avidity, though no other carnivorous animals will attack it. This," says Scaliger, still apostrophizing Cardano, "is merely a little sauce and seasoning to your allusion to the fable of the Lamb; but I would like to know from you how four distinct legs and their feet can be produced from one stem."

It is very remarkable that this dissertation of Scaliger, which is really a keen satire on Cardano, and a sarcastic repetition of his version of the fable with ironical comments thereon, has been almost invariably taken as serious, and regarded as an expression of his entire belief in the "Scythian Lamb," as described. Of all subsequent writers on the subject, Deusingius* seems to have been the only one who clearly perceived Scaliger's intention and meaning. Hence, many profound believers in the myth have claimed as their champion one who would have derided them for their credulity.

* Antonius Deusingius, Professor of Medicine, and Rector of the University of Groningen, in his '*Fasciculus Dissertationum Selectorum*,' p. 598, printed in 1660, declares his own utter disbelief in this animal-vegetable monstrosity, and after quoting Scaliger, thus writes of him :—"*Hæc equidem Scaliger, qui tamen ne serio historiam narrare credatur quam ipse revera pro fabulosa habet, nequaquam vero approbat, ut perperam de eo refert Sennert.*"—*Hyp. Physic.* 5, cap. 8.

Claude Duret, for example, whose implicit faith in the marvellous zoophyte nothing could shake, quotes verbatim in its defence the remarks of "le grand Jules César Scaliger," and asks [*] triumphantly,—

"Who cannot see plainly that Cardano, after having long doubted, and after having adduced philosophical arguments drawn from the works of Aristotle and other eminent writers, felt himself obliged and condemned to confess that in a place filled with heavy and dense air (such as is Tartary) the Borametz—true plant-animals—might exist as described, as well as sponges, 'sea-nettles,' and 'sea-lungs,' which every one knows are true zoophytes, or animal-plants."

After this amusing assumption that the air of Tartary possesses the "weight" and "density" necessary for the production of plant-animals, Duret quotes from Sir John Mandeville's book in the language in which it was originally written—the Romanic—the passage which I have extracted from the old English version of the enterprising knight's 'Voiage and Travailes,' and also cites, in confirmation of the prodigy, the account given of it by the Baron Von Herberstein. He then strongly expresses his own belief that—

"Of all the strange and marvellous trees, shrubs, plants and herbs which Nature, or, rather, God himself, has produced, or ever will produce in this Universe, there will never be seen anything so worthy of admiration and contemplation as these 'Borametz' of Scythia, or Tartary,—plants which are also animals, and which browze and eat as quadrupeds... If I did not entirely believe this I would denounce it as fabulous, instead of accepting it as a fact; but those who are in the habit of daily studying good and

---

[*] '*Histoire admirable des Plantes*,' p. 322.

rare books, printed and in manuscript, and who are endowed with great wisdom and understanding, know that there is no impossibility in Nature, *i.e.* God himself, to whom be all the honour and glory!"

Besides the authors already quoted, and others who merely copied the narratives of their predecessors, Guillaume de Saluste, the Sieur du Bartas, accepted as authentic the story of the Vegetable Lamb. In his poem "*La Semaine*," published in 1578, in which the first few days of the existence of all terrestrial things are described reverently and with considerable power, he represents this plant as one of those which excited the astonishment of the newly-created Adam as he wandered on the first day of the second week through the Garden of Eden, the earthly Paradise in which he had been placed.

"Or, confus, il se perd dans les tournoyements,
Embrouillées erreurs, courbez desvoyements,
Conduits virevoultez, et sentes desloyales
D'un Dedale infiny qui comprend cent Dedales,
Clos non de romarins dextrement cizelez
En hommes, my-chevaux, en courserots scelez,
En escaillez oyseaux, en balènes cornues,
Et mille autres façons de bestes incogneues,
Ains de vrays animaux en la terre plantez,
Humant l'air des poulmons, et d'herbes alimentez,
Tels que les Boramets, qui chez les Scythes naissent
D'une graine menues, et des plantes repaissent ;
Bien que du corps, des yeux, de la bouche, et du nez,
Ils semblent des moutons qui sont naguières naiz.
Ils le seroient du vray, si dans l'alme poictrine
De terre ils n'enfonçoient une vive raçine
Qui tient à leur nombril, et tombe le mente jour
Quils ont brouttè le foin qui croissoit à l'entour,
O merveilleux effect de dextre divine,
La plante a chair et sang, l'animal a raçine,
La plante comme en rond de soymême se meut,
L'animal a des pieds, et si marcher ne peut :

> La plante est sans rameaux, sans fruict, et sans feuillage,
> L'animal sans amour, sans sexe, et vif lignage ;
> La plante a belles dents, paist son ventre affamè
> Du fourrage voisin, l'animal est sémè."

Joshua Sylvester, the admiring translator of Du Bartas,[*] gives the following version of the above lines :—

> " Musing, anon through crooked walks he wanders,
> Round winding rings, and intricate meanders.
> False-guiding paths, doubtful, beguiling, strays,
> And right-wrong errors of an endless maze ;
> Nor simply hedged with a single border
> Of rosemary cut out with curious order
> In Satyrs, Centaurs, Whales, and half-men-horses,
> And thousand other counterfeited corses ;
> But with true beasts, fast in the ground still sticking
> Feeding on grass, and th' airy moisture licking,
> Such as those Borametz in Scythia bred
> Of slender seeds, and with green fodder fed ;
> Although their bodies, noses, mouths, and eyes,
> Of new-yeaned lambs have full the form and guise,
> And should be very lambs, save that for foot
> Within the ground they fix a living root
> Which at their navel grows, and dies that day
> That they have browzed the neighbouring grass away.
> Oh ! wondrous nature of God only good,
> The beast hath root, the plant hath flesh and blood.
> The nimble plant can turn it to and fro,
> The nummed beast can neither stir nor goe,
> The plant is leafless, branchless, void of fruit,
> The beast is lustless, sexless, fireless, mute :
> The plant with plants his hungry paunch doth feede,
> Th' admired beast is sowen a slender seed."

About the middle of the seventeenth century very little belief in the story of the "Scythian Lamb" remained amongst men of letters, although it continued to be a

---

[*] 'Du Bartas : His Divine Weekes and Workes, translated and dedicated to the King's most excellent Maiestie by Joshua Sylvester, London. 1584.'

Fig. 3.—Adam and Eve admiring the Plants in the Garden of Eden. The "Vegetable Lamb" in the background.

Fac-simile of the Frontispiece of Parkinson's "Paradisus."

subject of discussion and research for at least a hundred and fifty years later.

Athanasius Kircher, Professor of Mathematics at Avignon, who wrote* in 1641, after following the error of his predecessors of quoting Scaliger as a believer in the myth, says :—

"Some authors have regarded it as an animal, some as a plant; whilst others have classed it as a true zoophyte. In order not to multiply miracles, we assert that it is a plant. Though its form be that of a quadruped, and the juice beneath its woolly covering be blood which flows if an incision be made in its flesh, these things will not move us. It will be found to be a plant."

This unwavering prediction has been fulfilled. But the story had to pass through many vicissitudes of acceptance and disbelief before this decision of Kircher was unanimously admitted to be correct. It seems to have been the fate of this curious fable, through the whole period of its history, that no sooner has a ray of some author's common sense penetrated the mist of superstition by which it was surrrounded than it has been again befogged by the ignorant credulity of the next writer on the subject.

Jans Janszoon Strauss, a Dutchman, better known as Jean de Struys, who travelled through many countries, and amongst them Tartary, from 1647 to 1672, describes † this vegetable wonder. But he was an uneducated and credulous man, and his account of it is little more than a repetition of the errors and fallacies of former centuries concerning it, rendered still more incomprehensible by his

* 'Magnes; sive de arte magneticâ opus tripartitum,' p. 730.
† 'Voyages de Jean de Struys en Moscovie, en Tartarie, et en Perse,' chap. xii. p. 167. Amsterdam. 1681. Also an English translation, "done out of Dutch," by John Morrison. London. 1684. See Appendix E.

having confused with its "very white down, as soft as silk," the Astrachan lamb-skins, which were then, and are still, a well-known article of commerce. He says :—

"On the west side of the Volga is a great dry and waste heath, called the Step. On this heath is a strange kind of fruit found, called 'Baromez' or 'Barnitsch,' from the word 'Boran,' which is "a Lamb" in the Russian tongue, because of its form and appearance much resembling a sheep, having head, feet and tail. Its skin is covered with a down very white and as soft as silk. The Tartars hold this in great esteem, and it is sold for a high price. I have myself paid five or six roubles for one of these skins, and doubled my money when I sold it again. The greater number of persons have them in their houses, where I have seen many. That which caused me to observe it with greater attention was that I had seen one of these fruits among the curiosities in the house of the celebrated Mr. Swammerdam, in Amsterdam, whose museum is full of the rarest things in Nature from distant and foreign lands. This precious plant was given to him by a sailor who had been formerly a slave in China. He found it growing in a wood, and brought away sufficient of its skin to make an under-waist-coat. The description he gave of it did very much agree with what the inhabitants of Astrachan informed me of it. It grows upon a low stalk, about two and a half feet high, some higher, and is supported just at the navel. The head hangs down, as if it pastured or fed on the grass, and when the grass decays it perishes : but this I ever looked upon as ridiculous ; although when I suggested that the languishing of the plant might be caused by some temporary want of moisture, the people asseverated to me by many oaths that they have often, out of curiosity, made experience of that by cutting away the grass, upon which it instantly fades away. Certain it is that there is nothing which is more coveted by wolves than this, and the inward parts of it are more congeneric with the anatomy of a lamb than mandrakes are with men. However, what I might further say of this fruit, and what I believe of the wonderful operations of a secret sympathy in Nature, I shall rather keep to myself than aver, or impose upon the reader with many other things which I am sensible would appear incredible to those who had not seen them."

The next traveller, in order of date, who made the Tartarian Lamb the object of his investigations was Dr. Engelbrecht Kaempfer, who, in 1683, accompanied an embassy to Persia, and was appointed Surgeon to the Dutch East India Company two years later. He reported, on his return, that he had searched "*ad risum et nauseam*" for this "zoophyte feeding on grass," that there was nothing in the country where it was believed to grow that was called "Borametz," except the ordinary sheep, and that all accounts of a sheep growing upon a plant were mere fiction and fable. "The word 'Borametz,' he says,* "is a corruption of the Russian 'Boranetz,' in Polish 'Baranak,' the diminutive of which, 'Baran,' is Sclavonic. In such a case it signifies 'a sheep.' But," he continues, "there is in some of the provinces near the Caspian Sea a breed of sheep totally different from those with which we are commonly acquainted, and highly valued for the elegance of the skin, which is used in various articles of clothing by the Tartars and Persians. For the magnates and the rich who desire a material superior to that worn by the general population,

---

* '*Amœnitatum Exoticarum politico-physico-medicarum fasciculi*,' x., lib. 3, obs. 1. Lemgo, 1712. Kaempfer's MSS. and collections were acquired by Sir Hans Sloane, and were deposited in the British Museum.

the skins of the youngest lambs are preserved, the fleeces of these being much softer that those of the older ones, and the younger the animal from which they are taken the more costly are they." He then refers to the barbarous custom of killing the ewes before the time of natural parturition to obtain possession of the immature fleece of the unborn lamb, and says, correctly, that the earlier the stage of pregnancy in which this operation is performed the finer and softer is the fur of the fœtal skin, and the lighter and closer are the little curls for which it is chiefly prized. The pelt, also, is so thin that it is scarcely heavier than a membrane, and, in drying, it frequently shrinks so as to lose all similitude to the skin of a lamb, and assumes a form which might lead the ignorant and credulous to believe that it was a woolly gourd. He, therefore, conjectures that some of these dried and shrunken skins may have been placed in museums as examples of the fleece of the "Tartarian Lamb," under the supposition that they were of vegetable origin.

Kaempfer's suggestions were ingenious, though his theory was erroneous. But, although he rather impeded than assisted in the correct identification of the object of discussion, he, at least, helped to discredit the myth, which he declared to be one of those "received with favour by the superstitious, and which when once they have found a writer to describe them, however incorrectly, please the many, obtain numerous adherents, and become respectable by age."

An important chapter in the history of this curious fiction was reached when, in 1698, Sir Hans Sloane laid before the Royal Society an object which has ever since been generally regarded as a specimen of the strange natural

* Philosophical Transactions, vol. xx. p. 861 ; and Lowthorp's Abridgment of the Phil. Trans. vol. ii. p. 649.

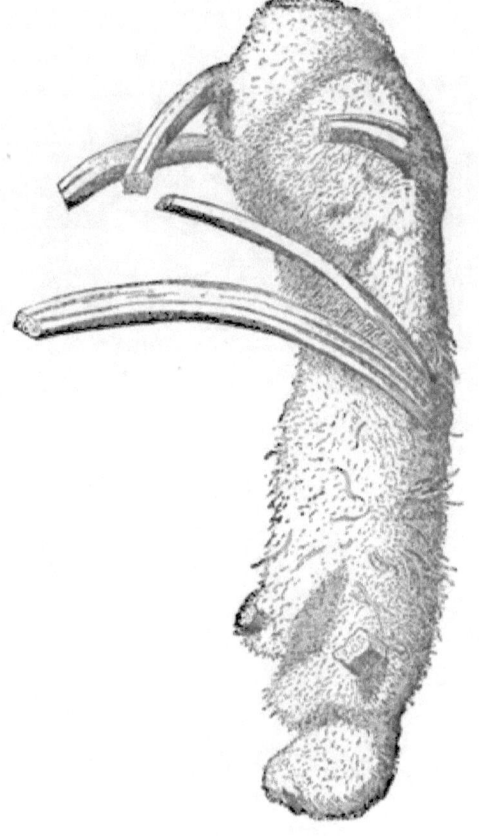

Fig. 4.—Rhizome of a fern, shaped by the Chinese to represent a tan-coloured dog, and laid before the Royal Society by Sir Hans Sloane as a specimen of the "Barometz," or "Tartarian Lamb."

From the 'Philosophical Transactions,' vol. xx. p. 861.

production about which so much mystery had existed, so many outrageous stories had been told, and on which so much learned discussion had been expended. His description of it is printed in the Society's Transactions, and is as follows :—

"The figure (fig. 4) represents what is commonly, but falsely, in India, called 'the Tartarian Lamb,' sent down from thence by Mr. Buckley.* This was more than a foot long, as big as one's wrist, having seven protuberances, and towards the end some foot-stalks about three or four inches long, exactly like the foot-stalks of ferns, both without and within. Most part of this was covered with a down of a dark yellowish snuff colour, some of it a quarter of an inch long. This down is commonly used for spitting of blood, about six grains going to a dose, and three doses pretended to cure such a hæmorrhage. In Jamaica are many scandent and tree ferns which grow to the bigness of trees, and have such a kind of *lanugo* on them, and some of the capillaries have something like it. It seemed to be shaped by art to imitate a lamb, the roots or climbing parts being made to resemble the body, and the extant foot-stalks the legs. This down is taken notice of by Dr. Merret at the latter end of Dr. Grew's Mus. Soc. Reg. by the name of 'Poco Sempic,' a 'golden moss,' and is there said to be a cordial. I have been assured by Mr. Brown, who has made very good observations in the East Indies, that he has been told by those who lived in China that this down or hair is used by them for the stopping of blood in fresh wounds, as cob-

---

* This specimen evidently came from China; for I find a record that at the date of Sir Hans Sloane's paper "Mr. Buckley, Chief Surgeon at Fort St. George, in the East Indies, presented to the Royal Society a cabinet containing Chinese surgical and other instruments and simples."

webs are with us, and that they have it in so great esteem that few houses are without it; but on trials I have made of it, though I may believe it innocent, yet I am sure it is not infallible."

Sir Hans Sloane had, it is true, clearly perceived the nature of the specimen sent to the Royal Society by Mr. Buckley, and had correctly identified it as a portion of one of the arborescent ferns; but on the question whether he had discovered the right interpretation of the puzzling enigma I shall have more to say presently. The object figured seems to have been regarded by many of his contemporaries as so insufficient to meet the requirements of the oft-told story of the plant-animal, and so unsatisfactory an explanation of it, that every one who subsequently had an opportunity of visiting Tartary still felt it to be his duty to make enquiries concerning the famous prodigy of that country.

Accordingly, we find that John Bell, of Autermony, availed himself of the opportunity afforded him by a diplomatic journey to Persia,[*] in 1715-1722, to endeavour, whilst in Tartary, to obtain authentic information respecting the "Vegetable Lamb." He found that nothing was known of it in the country where it was supposed to be indigenous, and thus writes of it :—

"Before I leave Astracan, it may be proper to rectify a mistaken opinion which I have observed to occur in grave German authors, who, in treating of the remarkable things of this country relate that there grows in this desart, or stepp adjoining to Astracan, in some plenty, a certain

[*] 'Travels from St. Petersburg in Russia to various parts of Asia, in 1716, 1719, 1722, &c., by John Bell, of Autermony. Dedicated to the Governor, Court of Assistants, and Freemen of the Russia Company. London. 1764.' See Appendix F.

shrub or plant called in the Russian language 'Tartasky Borashka,' *i.e.* 'Tartarian Lamb,' with the skins of which the caps of the Armenians, Persians, Tartars, &c., are faced. They also write that the 'Tartashky Borashka' partakes of animal, as well as vegetative life, and that it eats up and devours all the grass and weeds within its reach. Though it may be thought that an opinion so very absurd could find no credit with people of the meanest understanding, yet I have conversed with some who were much inclined to believe it, so very prevalent is the prodigious and absurd with some part of mankind. In search of this wonderful plant I walked many a mile accompanied by Tartars who inhabit these desarts; but all I could find out were some dry bushes, scattered here and there, which grow on a single stalk with a bushy top of a brownish colour: the stalk is about eighteen inches high, the top consisting of sharp prickly leaves. It is true that no grass or weeds grow within the circle of its shade—a property natural to many other plants, here and elsewhere. After a careful enquiry of the more sensible and experienced among the Tartars, I found they laughed at it as a ridiculous fable."

Bell further says:—

"In Astracan they have large quantities of lamb-skins, grey and black, some waved and others curled, all naturally and very pretty, having a fine gloss, especially the waved, which at a small distance appear like the richest watered tabby:* they are much esteemed, and are much used for the lining of coats and the turning up of caps, in Persia, Russia, and other parts. The best of these are brought from Bucharia, China, and the countries adjacent, and are taken from the ewe's belly after she hath been killed, or the

---

* A rich watered silk: from the French "*tabis*"; Italian, "*tabi*"; Persian, "*retabi*."

lamb is killed immediately after it is lambed, for such a skin is equal in value to the sheep. The Kalmuks and those Tartars who inhabit the desert in the neighbourhood of Astracan have also lamb-skins which are applied to the same purpose, but the wool of these being rougher and more hairy, they are inferior to those of Bucharia and China both in gloss and beauty, and also in the dressing; consequently in value. I have known one single lamb-skin from Bucharia sold for five or six shillings sterling, when one of these would not yield two shillings."

Bell had sufficient discrimination to see that these Astracan lamb-skins were in no way connected with the fable of the "Borametz," and thus avoided the error of Kaempfer, who regarded them as having given rise to the reports of the existence of that marvellous "animal-plant."

The Abbé Chappe-d'Auteroche, during his visit to Tartary,[*] about half a century later than John Bell, sought for the "Scythian Lamb" with equal earnestness and with similar want of success.

Long, however, before the result of the investigations of these two travellers had been made known, a second manipulated fern-root, similar to that described by Sir Hans Sloane, had been subjected to the scrutiny of another keen and scientific observer.

In September, 1725, Dr. John Philip Breyn, of Dantzic, addressed to the Royal Society of London an important communication in Latin on this subject,[†] in which he expressed his complete disbelief in the old story, and described a specimen of the "Borametz" (as he believed it

---

[*] 'Voyage en Sibérie,' Paris. 1768.

[†] '*Dissertiuncula de Agno Vegetabili Scythico, Borametz vulgo dicto.*' Phil. Trans., vol. xxxiii. p. 353, 1725; and also in Martyn's Abridgment of the Phil. Trans., vol. vi. p. 317.

Fig. 5.—Rough model of a tan-coloured dog, shaped by the Chinese from the rhizome of a fern, and submitted to the Royal Society by Dr. Breyn as a specimen of the "Scythian Vegetable Lamb."

From the 'Philosophical Transactions,' No. 390.

to be) which had fallen into his hands, and which had led him, independently, to the same conclusion as that arrived at by Sir Hans Sloane, of whose observations, he says, he was unaware when his own memoranda were written. Commencing by quoting the maxim, " *Non fingendum sed inveniendum quid Natura faciat aut ferat*," he urges upon all who search for the hidden treasures of Nature, or who desire to discover her secrets, to bear in mind that golden axiom that "the works and productions of Nature should be discovered, not invented," and remarks that, if the older writers had adhered to this, Natural History, great and honourable in itself, would not have been tarnished by so many silly fables like that of the "Scythian Lamb." He directs attention to the fact that none of those who have described this plant-animal are able to say that they ever saw it growing; quotes Kaempfer's interpretation of the origin of the report, namely the Astrachan lamb-skins of commerce, and hesitates to regard the object in his possession as the key of the problem. That he had grave and sufficient reasons for his doubts upon this point will be seen from his interesting description of the curiosity referred to. He says :—

" A certain learned and observant man, passing through our city on his return from a journey through Muscovy, enriched my museum with, amongst other natural curiosities, one of these 'Scythian Lambs,' which he declared to be the genuine Borametz. It was about six inches in length, and had a head, ears, and four legs. Its colour was that of iron-rust, and it was covered all over with a kind of down, like the fibres of silk-plush, except upon the ears and legs, which were bare, and were of a somewhat darker tawny hue. On careful examination of it, I discovered that it was not an animal production, nor yet a fruit, but either the thick creeping root, or the climbing stem, of some plant, which by obstetric art had acquired the form of a quadruped animal. For the four legs, which looked as if the feet had been cut off from them, were so many stalks which had supported leaves, as were also those which formed the ears, and which more nearly resembled horns. The fibres emerging from these, by which, like other plants, this root or stalk had conveyed nutriment, left no doubt upon this point. Close inspection also showed that one of the front legs had been artificially inserted, and that the head and neck were not of one continuous substance with the body, but had been very cleverly and neatly joined on to it. In fact, this root, or stem, had been skilfully manipulated into the form of a lamb in the same artful manner as the little figures of men, which, it was said, shrieked and dropped human blood when drawn from the ground, were formed from the roots of the mandragore and bryony."

Dr. Breyn added that there remained in his mind some doubt as to the plant from which this burlesque of nature and art was fabricated, until the similarity of its ferruginous silky fibres to those of some of the capillaries suggested the thought that it must be a portion of some exotic fern. As to the particular species to which it belonged he was unable to pronounce an authoritative opinion, but, hoping in the course of time to receive more certain information concerning it, he would merely say that he believed it was of a peculiar species found in Tartary, and up to that date undescribed.

Dr. Breyn's confirmation of Sir Hans Sloane's identification of the "Scythian Lamb" as the stem or rootlet of a fern artificially and cleverly manipulated was a crushing blow to the already weakened fable. Unfortunately, however, the conclusion thus arrived at was utterly misleading, though it not only satisfied his contemporaries, but has ever since—even to the present day—been universally accepted as the correct interpretation of the problem. The injurious result was, that, as the question appeared to have been set at rest, enquiry ceased, and for nearly sixty years afterwards no more was heard of the "Vegetable Lamb."

Towards the close of the century two eminent botanists, who were, of course, well acquainted with the specimens that had been described by Sir Hans Sloane and Dr. Breyn, were constrained in writing of the poetry of their science to make the legendary "Borametz" their theme.

Dr. Erasmus Darwin, in 1781, contributed to the literature of the subject the following lines*:—

> "E'en round the Pole the flames of love aspire,
> And icy bosoms feel the secret fire,
> Cradled in snow, and fanned by Arctic air,
> Shines, gentle Borametz, thy golden hair;
> Rooted in earth, each cloven foot descends,
> And round and round her flexile neck she bends,
> Crops the grey coral moss, and hoary thyme,
> Or laps with rosy tongue the melting rime;
> Eyes with mute tenderness her distant dam,
> And seems to bleat—a 'vegetable lamb.'"

Dr. Erasmus Darwin appears to have bestowed "golden hair" upon his Borametz, to assimilate it to the fern-root toys that were regarded as its prototypes; but as the fern of which they were made is a native of Southern China, and as no author has described the lamb-plant as being found in a cold climate, his authority and his motive for locating it in an arctic region are alike inexplicable.

Dr. De la Croix, the other botanical author above referred to, extolled, in 1791, the fabulous animal-plant

---

* 'The Botanic Garden.' A poem in two parts; with philosophical notes. London. 1781.

in a Latin poem* which Bishop Atterbury characterized as "excellent, and approaching very near to the versification of Virgil's 'Georgics.'"

"Qui Caspia sulcant
Æquora, sive legant spumosa Boristhenis ora
Sive petant Asiam velis, et Colchica regna,
Hinc atque inde stupent visu mirabile monstrum :
Surgit humo Borames. Præcelso in stipite fructus
Stat quadrupes. Olli vellus. Duo cornua fronte
Lanea, nec desunt oculi ; rudis accola credit
Esse animal, dormire die, vigilare per umbram,
Et circum exesis pasci radicitus herbis :
Carnibus Ambrosiæ sapor est, succique rubentes
Posthabeat quibus alma suum Burgundia Nectar ;
Atque loco si ferre pedem Natura dedisset,
Balatu si posset opem implorare voracis
Ora lupi contra, credas in stirpe sedere
Agnum equitem, gregibusque agnorum albescere colles."

As this has not been "done into English" (to use an old phrase), I venture to offer the following translation of it :—

"The traveller who ploughs the Caspian wave
For Asia bound, where foaming breakers lave
Borysthenes' wild shores, no sooner lands
Than gazing in astonishment he stands ;
For in his path he sees a monstrous birth,
The Borametz arises from the earth :
Upon a stalk is fixed a living brute,
A rooted plant bears quadruped for fruit,
It has a fleece, nor does it want for eyes,
And from its brows two woolly horns arise.
The rude and simple country people say
It is an animal that sleeps by day
And wakes at night, though rooted to the ground,
To feed on grass within its reach around.
The flavour of Ambrosia its flesh
Pervades ; and the red nectar, rich and fresh,
Which vineyards of fair Burgundy produce
Is less delicious than its ruddy juice.

* 'Connubia Florum, Latino Carmine Demonstrata.' Bath. 1791.

FIG. 6.—THE "BORAMETZ," OR "SCYTHIAN LAMB."

From De la Croix's 'Connubia Florum.'

The central figure is a copy of Zahn's picture of the fabulous plant-animal; the other two are taken from fern-root specimens supposed to be "Vegetable Lambs."

> If Nature had but on it feet bestowed,
> Or with a voice to bleat the lamb endowed,
> To cry for help against the threat'ning fangs
> Of hungry wolves; as on its stalk it hangs,
> Seated on horseback it might seem to ride,
> Whit'ning with thousands more the mountain side."

We must now leave the poetical view of the subject, and come to facts.

The substance of which the artificial animals exhibited by Sir Hans Sloane and Dr. Breyn were constructed is the long root-stock of a fern of the genus *Dicksonia*, of which there are from thirty to thirty-five species, varying greatly in size, in their mode of growth, and in the cutting of their fronds. Some of them, such as *D. antarctica*, a native of Australia and New Zealand, often seen in our greenhouses, are tree-like in habit, having stems from ten to forty feet in height, and fronds two or three yards in length, and two feet or more across; whilst others have root-stocks creeping along the surface of the ground. The genus is most fully represented in tropical America and Polynesia: one species extends as far north as the United States and Canada, and another was introduced into this country from St. Helena. In some species, such as *D. Molluccensis*, from Java, the stems are furnished with strong hooked prickles; in others they are densely clad at the base with a thick coat of yellow-brown hairs, which shine almost like burnished gold. The stems of *D. Sellowiana*, from tropical America, are so thickly clad with long fibrous hairs, changing to brown or nearly black, that it has been said they precisely resemble the thighs of the howling monkeys.*

---

* See 'European Ferns,' By James Britten, F.L.S.; with coloured illustrations from Nature, by Dr. Blair, F.L.S. Cassell. London.—A work full of information on the culture, classification, and history of ferns. I am indebted to it for many of the details here given of the economic value of ferns.

The species of *Dicksonia* which has been supposed to have given origin to the fable of the "Scythian Lamb" has, from that circumstance, received the name of *Barometz*. It was formerly known as *Cibotium glaucescens*. It was introduced into cultivation in conservatories in this country about the year 1830, and was shortly afterwards described as *Cibotium barometz*, but the genus *Cibotium* is now generally united with *Dicksonia*. Its long caudex, or root-stock, creeps over the surface of the ground in the same manner as that of the better known "Hare's-foot" fern, *Davallia Canariensis*, and this is covered with long silky hairs, or scales, which look something like wool when old and dry. These hairs or scales have been sometimes used as a styptic in Germany, and also, very commonly, in China, as related to Sir Hans Sloane by Dr. Brown. The similar hairs of other species of *Dicksonia*, natives of the Sandwich Islands, are exported to the extent of many thousands of pounds weight annually under the name of "Pulu," and are used in the stuffing of mattrasses, cushions, &c. The hairs of *D. culcita* are similarly utilised in Madeira. No more than two or three ounces of hair are yielded by each plant, and it is reckoned that about four years must elapse before another gathering can be obtained.

The rhizomes and stems of many ferns abound in starch, and have a commercial value, either as medicine or food. The soft mucilaginous pith of *Cyathea medullaris*, one of the large tree-ferns of New Zealand, was formerly eaten by the natives. It is of a reddish colour, and, when baked, acquires a somewhat pungent flavour. In New Zealand ferns seem to be in some repute for their edible properties, for the large scaly rhizomes of *Marattia fraxinea*, and those of another fern, *Pteris esculenta*, nearly allied to our common bracken, *P. aquilina*, are also eaten by the Maoris.

The natives bake them in ashes, peel them with their teeth, and eat them with meat, as we do bread; and sometimes pound them between stones, in order to extract the nutritious matter, the woody part being rejected as useless. In Nepaul, the rhizomes of *Nephrolepis tuberosa* are similarly prepared for food; and in New Caledonia the mucilaginous matter of *Cyathea vieillardii* is obtained from incisions made in the stem, or at the base of the fronds. The succulent fronds of the little water-fern, *Ceratopteris thalictroides*, are boiled and eaten as a vegetable by the poorer classes in the Indian Archipelago. The young shoots of the handsome tree-fern, *Angiopteris evecta*, are eaten in the Society Islands, and its large rhizome, which is in great part composed of mucilage, yields, when dried, a kind of flour. In the same islands the young fronds of *Helminthostachys limulata*, the "Balabala" of the Fiji Islands, are eaten in times of scarcity; and the soft scales covering the *stipes* of the fronds are used by the white settlers for stuffing pillows and cushions in preference to feathers, because they do not become heated, and are thus more comfortable in a sultry climate. In New South Wales, the thick rhizome of *Blechnum cartilagineum* is much eaten by the natives. It is first roasted and then beaten, so as to break away the woody fibre: it is said to taste like a waxy potato.

By skilful treatment the inhabitants of Southern China occasionally converted the thick root-stock of one of these tree-ferns, "*Dicksonia barometz*," into a rough semblance of a quadruped, which quadruped, by a foregone conclusion, was supposed to be a lamb. They removed entirely the fronds that grew upward from the rhizome, excepting four, and these four they trimmed down until only about four inches of each stalk was left. The object thus shaped being turned upside down, the root-stock represented the body of the animal, and was supported by the four inverted stalks of the fronds, as upon four legs. If the specimen had an insufficient number of stalks growing from it to make the four legs, others were artificially and neatly affixed to it; ears were similarly provided, and, if necessary, the trunk was fitted with a head and neck made from another root-stock.

So far, well! The identification of the material of which these imitations of four-legged animals were fashioned as the rhizome and frond-stalks of a tree-fern is complete, and perfectly satisfactory. But, having given to these root-stocks of tree-ferns the full benefit of an acknowledgment of the economic uses that have been made of them in various ways and in different localities, and having frankly stated the still accepted theory of their connection with the myth of the "Vegetable Lamb of Scythia," I have to express my very decided opinion that they and the "lambs" (?) made from them had no more to do with the origin of the fable of the "*Barometz*" than the artificial mermaids so cleverly made by the Japanese have had to do with the origin of the belief in fish-tailed human beings and divinities. In the first place, as we shall presently see, these manipulated ferns were not intended by those who fashioned them to resemble lambs at all. Secondly, if they had been intended to represent the lamb of the fable, they could have been, like the Japanese mermaids, only the outcome and illustration of the legend—not the objects which first gave rise to it. Neither the one nor the other of these counterfeit fabrications appears to have been ever common; and neither was certainly manufactured in sufficient numbers, nor distributed so abundantly and completely over the habitable globe, as to have laid the foundation of a myth which in the one case was universally believed,* and in the other attracted attention all over Europe and Western Asia, and also in Egypt. Very few of the Japanese artificial mermaids have been seen in this country, though they have been eagerly sought for, and the fern-"lambs" that have been brought to England may be counted on one's fingers.†

* See the Chapter on "Mermaids" by the Author in 'Sea Fables Explained,' one of the Handbooks issued by the Authorities of the Great International Fisheries Exhibition of 1883. London. Clowes and Sons, Limited.

† I know of only four—(though, of course, there may be others, of which I shall be glad to receive information)—namely, one in the Botanical department of the British Museum; another in the Museum of the Royal College of Surgeons; the specimen sent from India by Mr. Buckley to the Royal Society in 1698; and that described by Dr. Breyn in 1725. Of the origin of the first-mentioned nothing is known, though it is apparently the one figured by John and Andrew Rymsdyk, in their '*Museum Britannicum*' (1778, plate xv.), as one of the curious objects in the British Museum. Of the second we only know that it was presented to the College of Surgeons by Mr. Quekett—the habitat of the fern of which it is composed being erroneously given in the Catalogue (No. 177 of "Plants and Invertebrates") as "Plains of Tartary," the supposed home of the mythical lamb, but where the fern in question never grew. That sent to England by Mr. Buckley, and which was the subject of Sir Hans Sloane's paper in 1698, seems to have been lost or mislaid. Whether it remained in the possession of the Royal Society, or was placed by Sir Hans Sloane in his own collection, it ought to be in the British Museum. But nothing is known of it there, nor of the cabinet of surgical instruments and appliances in which it arrived. I have endeavoured to trace it; but although, as usual, I have met with every kind assistance and courtesy from the heads of departments, I have been unsuccessful.

Sir Hans Sloane, who died in 1753, bequeathed his valuable collection and library to the nation on the condition that £20,000 should be paid to his executors for the benefit of his daughters. The Government raised the necessary funds by a guinea lottery, and sufficient money was thus obtained to purchase also (for £10,500) Montague House, in Bloomsbury, which then became the British Museum.

Further, it is a fact which seems to have been strangely overlooked, that these tree-ferns, with the creeping root-stocks, do not grow in Tartary. The particular species of *Dicksonia* from which the doll-"lambs" were made is a native of Southern China, Assam, and the Malayan peninsula and islands.* And we have conclusive evidence, in addition to the report made by Mr. Buckley to the Royal Society (p. 27), that these playthings themselves were of Chinese workmanship.

Juan de Loureiro, an accomplished Portuguese botanist and Fellow of the Royal Society of Lisbon, who lived and laboured as a Catholic missionary for more than thirty years in Cochin China, and, afterwards, for three years in China, thus writes † :—

"The *Polypodium borametz* grows in hilly woods in China and Cochin China. Many authors have written of the Scythian Lamb, or Borametz—most of them fabulously. Ours is not a fruit, but a root, which is easily shaped by the help of a little art into the form of *a small rufous dog, by which name, and not by that of a 'lamb,' it is called by the Chinese.*"

When the Royal Society removed from their old premises, in Crane Court, to Somerset House in 1780 they also gave the contents of their cabinets to the National Collection, but many of these, and amongst them this fern-root animal, cannot be found.

Dr. Breyn, of Dantzic, no doubt retained the specimen which he described, and it is probably in some continental collection.

I know, therefore, of only two of these so-called "lambs" extant in this country—one in the Natural History Museum, South Kensington, and the other in the Museum of the Royal College of Surgeons. No history of either of these has been preserved.

* '*Synopsis Filicum*,' by Sir W. J. Hooker and J. G. Baker, F.L.S. 1863. Art. "Dicksonia barometz."

† *Flora Cochinchinensis*, tom. i. p. 675. Lisbon. 1790.

Loureiro describes the cutting off the stalks to form the legs, the fixing on of smaller ones as ears, and other particulars of the rude manufacture of these fern-root dogs, as witnessed by himself. The common name of these toys in China—" Cau-tich," and in Cochin China, " Kew-tsie," both represent a "tan-coloured dog."

It must also be borne in mind that the lamb-plant was represented as springing from a seed like that of a melon, but rounder, and that the natives of the country where it grew planted these seeds. It was therefore a cultivated plant. The lamb, it was also stated, was contained within the fruit or seed-capsule of the plant; and when this fruit, or seed-pod, was ripe it burst open, and the little lamb within it was disclosed. The wool of this lamb was described by various writers as being "very white," "as white as snow," whereas these root-stocks of ferns bear no resemblance to a lamb in their natural condition; and when they have been deftly trimmed into shape the hairs or scales upon them are tawny orange, matching better with the "tan" markings of a dog, which they were intended to represent, than with the soft, white fleece of a young lamb.

Therefore, even if I had no better explanation to offer, I should be led to the conclusion that the identification of these *tawny* toy-*dogs*, made in *China* from the *root* of a *wild* fern, the spores of which are *as small as dust*, with the "Vegetable *Lambs*" of *Scythia*, whose *white* fleeces were found within the ripe and opening *fruit* of a *cultivated* plant, raised from *a large seed*, was obviously erroneous, and that the origin of the rumour must be sought for elsewere.

The plant that set all Europe talking of the lambs that grew in fruits and on stalks of plants somewhere in Scythia was one of far higher importance and value to mankind than the childish knick-knacks made for amusement out of the creeping root-stocks of ferns. These and the curly-fleeced progeny of the poor ewes of Astrachan were lambs that crossed the track of the first, lost lamb, and led those searching for it into the mistake of following their respective trails, whilst the original "Scythian Lamb" escaped from sight.

Tracing the growth and transition of this story of the lamb-plant from a truthful rumour of a curious fact into a detailed history of an absurd fiction, I have no doubt whatever that it originated in early descriptions of the cotton plant, and the introduction of cotton from India into Western Asia and the adjoining parts of Eastern Europe.

Herodotus, writing (B.C. 445) of the usages of the people of India, says (lib. iii. cap. 106) of this cotton :—"Certain trees bear for their fruit fleeces surpassing those of sheep in beauty and excellence, and the natives clothe themselves in cloths made therefrom."

In the 47th chapter of the same book, Herodotus describes a corselet sent by Aahmes (or Amasis) II., King of Egypt, to Sparta as having been "ornamented with gold and *fleeces from the trees*"—padded with cotton, in fact.

Ctesias, also, who was the contemporary of Herodotus, and was made prisoner, and kept by the King of Persia as his court physician for seventeen years, was acquainted with the use of a kind of wool, the produce of trees, for spinning and weaving amongst the natives of India, for he mentions in his '*Indica*' a fragment quoted by Photius, "tree-garments"; and that he thus referred to clothing made from these tree-fleeces we have the testimony of Varro :— "Ctesias says that there are in India *trees that bear wool*."

Nearchus, the admiral of Alexander the Great, reported that "there were in India trees bearing, as it were, flocks or bunches of wool, and that the natives made of this wool garments of surpassing whiteness, or else their black complexions made the material appear whiter than any other."

Aristobulus, another of Alexander's generals, made mention in his journal of the cotton plant, under the name of "the wool-bearing tree," and stated that "it bore a capsule that contained seeds which were taken out, and that which remained was carded like wool."

Strabo, who records this (lib. xv. cap. 21), referring to it in another paragraph, writes :—" Nearchus says that their (the natives') fine clothing was made from this wool, and that the Macedonians used it for mattresses and the stuffing of their saddles."*

Theophrastus, the disciple of Aristotle, writing about B.C. 306, says † :—

"The trees from which the Indians make their clothes have leaves like those of the black mulberry, but the whole plant resembles the dog-rose. They are planted in rows on the plains, so as to look like vines at a distance."

In another passage of the same book (cap. 9) he writes :—

"In the Island of Tylos, which is in the Arabian Gulf,‡

---

* Unfortunately the Journal and Narrative of Nearchus, written B.C. 325-324, are lost, as are also those of Aristobulus, who seems to have been a very accurate observer ; and we are indebted to Strabo and Arrian for the summaries and extracts from them that we possess. Strabo's *'Geographia'* was completed A.D. 21, about three years before his death. Fabius Arrianus wrote his *'Historia Indica,'* and *'Periplus Maris Erythræi,'* which contain valuable particulars of Alexander's expedition, about A.D. 131-135.

† *'De Historia Plantarum,'* lib. iv. cap. 4.

‡ Theophrastus is in error in placing Tylos in the Arabian Gulf (which we now call the Red Sea) ; it was in the Persian Gulf, and is now known as Bahrsin. The ancients, however, gave to the whole of

the wool-bearing trees, which grow there abundantly, have leaves like the vine, but smaller. They bear no fruit, but the pod containing the wool is about the size of a spring apple ("μῆλον"), whilst it is unripe and closed, but when it is ripe it opens: the wool is then gathered from it, and woven into cloths of various qualities—some inferior, but others of great value."

This description by Theophrastus is remarkably correct as applied to the herbaceous variety of the cotton-plant, from which the chief supply of cotton for spinning and weaving into cloth has always been obtained. In its mode of growth—branched, spreading, and flexible—it may well be likened to the dog-rose; and its palmate leaves bear a close resemblance to those of the black mulberry, which differ little from the leaves of some varieties of the vine. The remark relative to the mode of cultivation is also exactly applicable to the cotton-plant, which is set in rows about four feet asunder, and the plants about two feet apart, so that a field of it resembles a vineyard when seen from a distance.

Pomponius Mela, the author next in order of time, also writes in his account of India * of the "trees that produce wool used by the natives for clothing."

Then comes Pliny, who, incompetent and worthless as a naturalist, though admirable as a writer, obscured this subject, as he did many others. In his 'Natural History'†

---

the sea between the east coast of Africa, north of Mogador, and the west shores of India the name of the "Erythræan Sea," from King Erythros, of whom nothing more is known than the name, which, in Greek, signifies "red." From this casual meaning of the word it came to be believed that the water of this sea differed in colour from that of others, and that it was consequently more difficult to navigate.

* *De Situ Orbis*, lib. iii. cap. 7.
† '*Naturalis Historia*,' A.D. 77.

he mentions cotton in four different paragraphs, and in every one of them inaccurately. He confuses cotton with flax, and the fabrics woven of it with linen, and treats of silk as a downy substance scraped from the leaves of trees. And, in transcribing, or translating, the passage from Theophrastus relating to the "wool-bearing trees," he distorts the author's words, and states that "these trees bear *gourds* the size of a quince, which burst when ripe, and display balls of wool out of which the inhabitants make cloths like valuable linen." Pliny therefore seems to have been the author of the "gourd" portion of the story which afterwards obtained currency in Western Europe.

I shall quote one more ancient mention of the "fleece-bearing plant," because the author of it gives a more exact description than any previous writer of that portion of it from which the wool is taken.

Julius Pollux, who wrote about a hundred years later than Pliny, says in his 'Onomasticon':—

"There are also *Byssina* and *Byssus*, a kind of flax. But among the Indians a sort of wool is obtained from a tree. The cloth made from this wool may be compared with linen, except that it is thicker. The tree produces a fruit most nearly resembling a walnut, but three-cleft. After the outer covering, which is like a walnut, has divided and become dry, the substance resembling wool is extracted, and is used in the manufacture of cloth."

This remark, of the pericarp of the cotton-pod, in some species of *Gossypium*, being three-cleft, is in accordance with fact, and is not noticed by any previous writer.

In tracing the development of these early and truthful accounts of the cotton-plant into the complete fable of the compound plant-animal, the "Vegetable Lamb of Scythia," we shall find it, as in the case of some other myths of the Middle Ages, attributable to two principal causes :—

1. The misinterpretation of ambiguous or figurative language ; 2. The similarity of appearance of two actually different and incongruous objects.

It is a curious fact, which I believe has not hitherto been noticed in connection with this subject, that the Greek word "μῆλον" (melon), very fitly used by Theophrastus in the passage quoted (p. 48) to describe the form and appearance of the unripe cotton-pod, may be equally correctly translated "a fruit," "an apple," or "a sheep" : the adjective "ἐαρινόν," which is also used, means "vernal" ; therefore the phrase may be regarded as signifying either that the vegetable wool was taken from a "spring apple" growing upon a tree, or from a "spring-sheep" (or lamb) growing upon a tree. Although I believe that the mistake originated, as I shall presently explain, in the actual and substantial resemblance between cotton wool and lamb's wool, rather than in the verbal identity of an appellative noun, it is not improbable that this ambiguous phrase of convertible interpretation may, in some measure, have contributed to convey, many centuries later, to readers of a dead language who knew nothing of the plant referred to, an erroneous idea of the nature of "the fleeces that grew on trees." It would seem so much more likely that a soft fleece of white wool should grow upon a young lamb yeaned in spring-time than inside a fruit like an apple in the partly-formed and unripe condition in which it is found in spring, that students in the Middle Ages, as they pondered doubtfully over this word of double meaning, would probably prefer the first interpretation, and translate the passage of Theophrastus as a statement that the wool was taken from a "spring-sheep," or lamb, growing upon a tree which bore no other fruit. It is also probable that this use of the Greek word "*melon*" gave rise to the report in later times that the seed of the plant which bore the "Vegetable Lamb" was like that of a melon or gourd.

We may next take into account the prevalence amongst many tribes and nations in both hemispheres of the custom of using figurative language in relation to the objects and occurrences of their daily life.

A very striking and remarkable proof is given us by Herodotus that the Scythians of the North-West, who carried both the cotton and the rumour of the lamb-plant into Muscovy, were in the habit of speaking thus figuratively and metaphorically. He writes (lib. iv. cap. 2) :—

"The part beyond the north, the Scythians say, can neither be seen nor passed through, by reason of the feathers shed there; for the earth and air are full of feathers, and it is these which interrupt the view."

Further on (lib. iv. cap. 31) he also observes :—

"With respect to the feathers with which the Scythians say the air is filled, and on account of which it is not possible either to see further upon the continent, or to pass through it, I entertain the following opinion. In the upper parts of this country it continually snows—less in summer than in winter, as is reasonable. Now, whoever has seen snow falling thick near him will know what I mean; for snow is like feathers, and on account of the winter being so severe the northern parts of this country are uninhabited. I think, then, that the Scythians and their neighbours call the snow feathers, comparing them together."

Herodotus was, of course, right in this interpretation.

Who can doubt that the people who would thus realistically describe snow as feathers would probably describe the white wool of the cotton-pod as "tree-lamb's-wool," the produce of a "lamb-plant," or "plant-lamb"?

The growth and development of the story of "the Scythian Lamb" from the similarity of appearance of two really different objects may be best explained by comparing it with another Natural-history myth, which ran curiously parallel with it. I allude to the fable that Sir John Mandeville tells us he related to his Tartar acquaintances, viz. that of the "*Barnacle Geese*"—which has never been surpassed as a specimen of ignorant credulity and persistent error.

From the twelfth to the end of the seventeenth century it was implicitly and almost universally believed that in the Western Islands of Scotland certain geese, of which the nesting-places were never found, instead of being hatched from eggs, like other birds, were bred from "shell-fish" which grew on trees. Upon the shores where these geese abounded, pieces of timber and old trunks of trees covered with barnacles were often seen which had been stranded by the sea. From between the partly opened shells of the barnacles protruded their plumose cirrhi, which in some degree resemble the feathers of a bird. Hence arose the belief that they contained real birds. The fishermen persuaded themselves that these birds within the shells were the geese whose origin they had been previously unable to discover, and that they were thus bred, instead of being hatched, like other birds, from eggs. As the tale spread to a distance, it gained by repetition, like the story of "The Three Black Crows" amusingly told by Dr. John Byrom.* The trees found upon the shore were soon reported to be trees growing on the shore; that which grew on trees people soon asserted to be the fruit of trees;

* See Appendix G.

and thus, from step to step, the story increased in wonder and obtained credit. It was discussed during many centuries by philosophers and men of learning, who, one after another, accepted the evidence in its favour, until Sir Robert Moray, F.R.S., in 1678, reported to the Royal Society that he had examined these barnacles, and that in every shell that he had opened he had "found a little bird—the little bill, like that of a goose; the eyes marked; the head, neck, breast, wings, tail, and feet formed, the feathers everywhere perfectly shaped and blackish-coloured, and the feet like those of other water-fowl." This nonsense was published in the 'Philosophical Transactions' (No. 137, January and February, 1678) under the auspices of the highest representatives of science in this country. The old botanist Gerard had previously (in 1597) had the audacity to assert that he had witnessed the transformation of the "shell-fish" into geese.*

In like manner the "wool-bearing plant" of Ctesias, Nearchus, Aristobulus, and Theophrastus, the plant of which Herodotus wrote that "it bore as its fruit fleeces which surpassed those of lambs in beauty and excellence," was soon reported to be "a plant bearing fruit within which was a little lamb having a fleece of surpassing beauty and excellence." As it was evident that a living lamb must take food, the "lytylle best" was, in the next version, kindly placed upon a stalk, and so balanced thereon as to be able to bend downward, and browze upon the surrounding herbage. Of course the lamb, if it fed on grass, must have digestive and other organs, like those of lambs ordinarily begotten, so these were liberally bestowed upon it with as much particularity as that exercised by Sir

* 'See 'Sea Fables Explained,' by the Author, 2nd edition, p. 114. Clowes and Sons, Limited.

Robert Moray in enumerating the "parts and features" of the "little tree-bird."* The transformation of the wondrous "plant-animal" from "a little lamb with a white fleece disclosed by the bursting of a ripe seed-pod growing on a stalk" into "a lamb growing on a stalk attached to its navel, and browzing on the herbage within its reach," vastly increased the difficulty of identifying it. Like the barnacle geese, it was discussed by philosophers and sought for by travellers; but its features had been distorted beyond recognition, and, instead of endeavouring to find its original portrait in the pages of old historians and geographers, enquirers looked for fresh information concerning it in the misleading tales of successive travellers. At last, as we have seen, another "vegetable lamb" crossed the trail of the original lost one, in the shape of the two Chinese toy-dogs laid before the Royal Society by Sir Hans Sloane and Dr. Breyn. That distinguished body of savants unfortunately accorded their recognition to the wrongful claimant, and ever since then botanists and antiquarians have regarded the problem as solved, and have been satisfied that in these few rude models of "tan-coloured dogs" they have found the true and original "snow-white" "Vegetable Lamb of Scythia."

The contented acceptance by botanists and other representatives of science, down to the present day, of three or four trumpery toys artificially and roughly fashioned by

---

* The figures of the ancient partly human, partly piscine deities, from which originated the belief in mermaids, similarly passed through various mutations. The first idea was that of a man coming out of the mouth of a fish. Subsequently, the form was that of a man clad in the skin of a fish—wearing it as a mantle—the head of the fish covering that of the man, like a cap or helmet. And so on, till a being was developed the upper half of whose body was human, and the lower half, from the waist downwards, that of a fish.

the Chinese from the rhizomes of a fern which does not grow in Tartary or Scythia, and brought to Europe by travellers at rare intervals, as sufficient to account for the origin of a rumour which spread from Asia all over Europe and attracted the attention of learned men of all countries for many centuries, is not the least remarkable circumstance in the history of the legend of the "Scythian Lamb."

Well might the old historians consider worthy of record the reports they had heard of the existence of the "wool-bearing tree," for, as Dr. Ure has remarked,* "it would be universally regarded as a miracle of vegetation did not familiarity blunt the moral feelings of mankind. This class of plants, largely distributed over the torrid zone, affords to the inhabitants a spontaneous and inexhaustible supply of the clothing material best adapted to screen their swarthy bodies from the scorching sun, and to favour the cooling influence of the breeze, as well as cutaneous exhalation. While the tropical heats change the soft wool of the sheep into a harsh, scanty hair, unfit for clothing purposes, they cherish and ripen the vegetable wool, with its more slender and porous fibres, admirably suited for clothing in a hot climate, as the grosser and warmer animal fibres are in a cold one. No sooner does the cotton pod arrive at maturity than its swollen capsules burst with an elastic force, in gaping segments, in order, as it were, to display to the most careless eye their white fleecy treasure, and to invite the hand of the observer to pluck it from the seeds, and to work it up into a light and beautiful robe. Thus held forth from the extremity of every bough, by its resemblance to sheep's wool it could not fail to attract attention."

Such keen observers as the ancient conquerors of India

* 'The Cotton Manufacture of Great Britain,' p. 71.

would have been sure to notice with surprise and interest the wonderful vegetable product which could be compared to nothing so aptly as to the white, soft wool of a little lamb, to appreciate its value and usefulness, and to admire the fabrics manufactured from it. And, as these fabrics gradually found their way northward from India by the great caravan routes, either by Samarcand, or by the passes of the Hindu Kush, by Bokhara and Khiva, through Turkestan and Tartary into Russia, in one direction, and by Egypt to the countries on the Mediterranean in another, the sensation they would cause is not difficult to realise. We can imagine how the newly-arrived trader, as he displayed his goods, would be eagerly questioned by intending purchasers of the novel, soft, white or coloured cloths, so well suited to their requirements, as to the nature of the raw material of which they had been woven. We can picture to ourselves their astonishment when he explained to them that the delicate, white, flossy fibres from which his fabrics were made, of which he, perhaps, showed them a sample, and which looked so like lamb's wool, was the produce of a plant, the fruit of which burst open when it became ripe, and exposed to view the white wool within it. And we can easily understand how the fame of this spread, and was carried into distant lands, and how this "vegetable lamb's wool" was discussed and talked about in countries where it, and the yarn spun from it, and the cloths woven from it, had not yet penetrated.

Now, let us complete our identification of the cotton-pod of India as "the Vegetable Lamb" of the fable by showing its right to the title of "the *Scythian* Lamb."

There is probably no race of men, or rather aggregate of races, mentioned prominently in history, of whom, and of whose country so little has been definitely known as of the ancient Scythians. They have been generally and vaguely, and, to a certain extent, correctly, regarded as represented in modern times by the numerous hordes of Tartars inhabiting the lands north of the mountains of the Caucasus, and part of central and northern Asia. So exclusively have they been identified with these tribes that the terms Tartary and Scythia have been looked upon as synonymous, and thus "the Scythian Lamb" has been called also the "Tartarian Lamb," or "the Vegetable Lamb of Tartary."

Under the name of "Scythia" was included (as may be seen on any good classical map) a vast territory, partly in Europe and partly in Asia, extending from the 25th to the 116th degree of East longitude. The European portion of it was comparatively a small province, known as "Scythia Parva," and comprised those districts of Silistria and Bessarabia bordering the western shores of the Black Sea, south of the mouths of the Danube. Scythia in Asia, which was separated from Scythia Parva by the two Sarmatias, included the whole of Turkestan, Thibet, Mongolia, and Siberia. It was bounded on the West by the Ural Mountains and river, and extended northward through then unknown regions to the Arctic Circle, and southward to the Himalayas. But still further south, beyond the western Himalayas—the Hindu-Kush—was another part of Scythia, known as "Indo-Scythia." This stretched southward to the Erythrean Sea (the Arabian Sea), and was that part of India now called Scinde and the Punjab. Through it flowed the Indus and the Hydaspes, and it was on the banks of the latter river, at Bucephalia (either the present Jhelum, or Jubalpore, eighteen miles lower), that Alexander's admiral collected the flotilla which he conducted down the Hydaspes to its confluence with the Indus, and along the whole course of that great river, and made his way by its lower mouth into the open water of the Arabian Sea. Then and there it was—from the time of their arrival in the country, during the war with Pontus and other Indian princes, and on their ten months' voyage homeward —that Alexander and his commodore Nearchus saw the native population of Indo-Scythia "clad in garments the material of which was whiter than any other, or at any rate appeared so in contrast with their wearers' swarthy skin," and which were "made of the wool like that of lambs, which grew in tufts and bunches upon trees."

Although more than two thousand years have passed since then, Nearchus's description of this costume—"a shirt, or tunic, reaching to the middle of the leg, a sheet folded about the shoulders, and a turban rolled round the head" —would be almost equally accurate at the present day. Its wearers may be congratulated that fashion has left unchanged and unspoiled an apparel so serviceable and well-suited to the climate of the country and the habits of its people!

As the "fleeces of vegetable wool, softer and whiter than that of the lamb," came from Indo-Scythia, the supposed plant-animal that bore them was first called "the Scythian Lamb."

As time passed on, the name of Scythia in Asia became merged in that of Tartary. From the time that the Mahometans became masters of Egypt and Constantinople, as no Christian was allowed to pass through their dominion to the East, intercourse with India by the two most direct roads ceased entirely. Cotton goods and other merchandise from India were therefore conveyed by the trading caravans before mentioned. The depôt to which they were generally forwarded was Samarcand, as was correctly related to Guillaume Postel by Michel, the Arabic interpreter (p. 13). There they met the great caravan travelling from the East into Russia, and, on the journey, passed through part of Scythia in Asia. In each district the caravan was joined by hosts of Tartar traders carrying with them the wool of their sheep, the hair of their goats, and the skins of both, the soft, curly skins of their lambs, and droves of hardy colts, the produce of their mares, whose milk was, and still is, to them as important an article of diet as that of cows is to ourselves. As the Tartar merchants brought with the fleeces of their sheep, goats, and lambs the fleeces also of "the fine white wool that grew on trees" and the piece-goods made from it, "the vegetable lamb" from which it was supposed to have been sheared became also in this manner identified with Tartary, in the same way as were Indian spices with "Araby," through which they sometimes passed in transit, but where they never grew. It thus became known as "the wool of the Tartarian Lamb," and travellers whose curiosity concerning the far-famed "zoophyte" was subsequently aroused sought for it in the dominions of the "Great Cham." But, just as when Æneas Sylvius Piccolomini, afterwards Pope Pius II., sought in Scotland for the "goose-bearing tree," which he eagerly desired to see, upon being told that it grew much further north, complained that "miracles will always flee farther and farther away"; so when any painstaking traveller in Tartary endeavoured to investigate the subject of the strange "plant-animal," he was sure to learn (unless he allowed himself to be cunningly hoaxed by the skin of a natural lamb, or the fruit of another plant) that the object of his search was non-existent in its reputed birthplace, and that he must look for it elsewhere.

Thus the story of the "Scythian" or "Tartarian Lamb"

grew, and was exaggerated and distorted, until all traces of its origin were so obliterated that even men of thought and learning have been unable to recognise in the misleading descriptions given of it the plant which, excepting corn, is, perhaps, the most valuable to mankind. For, as I have said, it seems to me to be clear and indubitable that the fruit which burst when ripe and disclosed within it "a little lamb" was the cotton pod, and that the soft, white, delicate fleece of "the Vegetable Lamb of Scythia" was that which we still call "Cotton Wool."

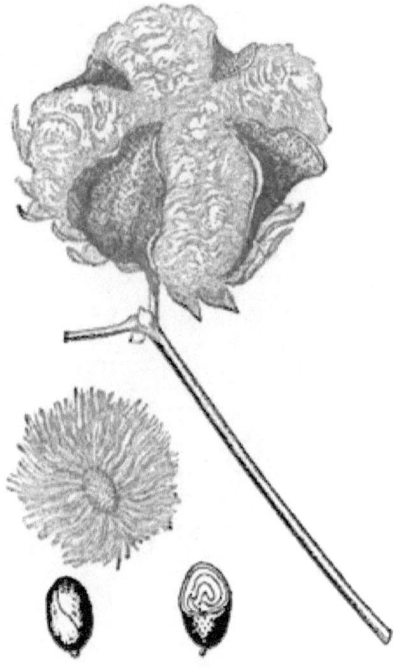

FIG. 7.—THE REAL "VEGETABLE LAMB"—A COTTON POD.
(*Gossypium herbaceum.*)

CHAPTER II.

THE HISTORY OF COTTON AND ITS INTRODUCTION INTO EUROPE.

IN the preceding pages I have referred to the introduction of cotton into the countries north and west of the Indus in so far only as the expressions of old writers relating to it have seemed to afford a clue to the origin of the fable of "the Scythian Lamb." But I venture to think that a brief account of its botanical affinities, and of its spread and distribution amongst various nations, may form an appropriate and acceptable sequel to the story of the wild rumours that preceded by many centuries its arrival in Western Europe.

The cotton plant, *Gossypium*, is one of the *Malvaceæ*—allied to the mallow. There are several varieties of it, but only three principal distinctions require notice—namely, the herbaceous, the tree, and the shrub species. The first and most useful, *Gossypium herbaceum*, is an annual plant, cultivated in the United States, India, China, and other countries. It grows to a height of from eighteen to twenty inches, and has leaves, which being somewhat lobed, of a bright dark green colour, and marked with brownish veins, were not inaptly compared by Theophrastus with those of the black mulberry and the vine. Its blossoms expand into a pale yellow flower, and when this falls off a three-celled, triangular capsular pod appears. The pod increases to the size of a large cob-nut or small medlar, and becomes brown as the woolly fruit ripens. The expansion of the wool then causes the pod to burst, and it discloses a ball of snow-white (in some species, yellowish) down consisting of three locks—one in each cell—enclosing and firmly adhering to the seeds. As the pods ripen the cotton is gathered by hand, and is exposed to the sun till it is perfectly dry; the seeds are then separated from it, and it is packed into bales for future use or exportation. In the United States it is planted in rows, four feet asunder, and the seeds are set in holes eighteen inches apart.

The shrub cotton grows in almost every country where the annual herbaceous cotton is found. Its duration varies according to the climate. In some places, as in the West Indies, it is biennial or triennial; in others, as in India, Egypt, &c., it lasts from six to ten years; in the hottest climates it is perennial; and in the cooler countries it becomes an annual.

The tree-cotton, *Gossypium arboreum*, grows in India, Egypt, China, the interior and western coast of Africa, and in some parts of America. As the tree only attains to a height of from twelve to twenty feet, it is difficult to distinguish the tree cotton and the shrub cotton when referred to by travellers.

The cotton plant, in all its varieties, requires a sandy soil. It flourishes on the rocky hills of Hindostan, Africa, and the West Indies, and will grow where the soil is too poor to produce any other valuable crop.

Cotton has always been regarded as indigenous to India, and as the characteristic clothing material of that country, as flax is of Egypt, silk of China, and the wool of sheep and goats of Northern Asia.

The uncertain nature of Hindoo chronology prevents our

even guessing at the period when cotton was first spun and woven in India; but there is little doubt that it was so used from the earliest ages of Hindoo civilization. As Dr. Robertson remarks, in his 'Historical Disquisition on British India'—"Whoever attempts to trace the operations of men in remote times, and to mark the various steps of their progress in any line of exertion, will soon have the mortification to find that the period of authentic history is extremely limited, and if we push our enquiries beyond the period when written history commences we enter upon the region of conjecture, of fable, and of uncertainty."

The earliest mention of cotton with which we are acquainted is found, according to Dr. Royle,* in the first book of the Rig Veda, Hymn 105, verse 8, which is supposed to have been composed fifteen centuries before the Christian era. It is, however, a mere allusion to "threads in the loom," and although it probably does refer to cotton, the evidence of this is only circumstantial. But in 'The Sacred Institutes of Manu,' which date from 800 B.C., cotton is referred to so repeatedly as to imply that it was in common use at that time in India. Dr. Royle says, on the authority of Professor Wilson, that cotton and cotton-cloth are mentioned in that book by the Sanscrit names "*Kurpasa*" and "*Karpasum*," and cotton-seeds as "*Kurpas-asthi*." The common Bengali name "Kupas," indicating cotton with the seed, which is still in general use all over India, and may even be occasionally heard in Lancashire, is, no doubt, derived from the Sanscrit, from which also comes the Latin "*carbasus*."

It is evident that the manufacture of cotton in India

* 'On the Culture and Commerce of Cotton in India and elsewhere,' by J. Forbes Royle, M.D., F.R.S. London. 1851.

must date from a very remote period indeed, for long before the time of Herodotus the processes of weaving and dyeing it had attained to a degree of excellence which indicates considerable previous experience; and a large export trade in white and coloured cotton fabrics had even then been established.

From India manufactured cotton seems to have reached Persia in very early times, for the word "Karpas" occurs in the book of Esther (chap. i. v. 6), in the description of the decorations of the palace of Shushan during the right royal festivities given there by King Ahasuerus, B.C. 519. In the verse referred to we are told that there were "white, green, and blue hangings." The word corresponding with "green" in the Hebrew is "*Karpas*," in the Septuagint and Vulgate, *carbasinus*, and should be rendered "cotton-cloth"; so that the hangings of the palace of Ahasuerus were of white and blue striped cotton, such as may be seen throughout India at the present day. Bishop Heber describes the Hall of Audience of the Emperor of Delhi, as having these striped curtains hanging in festoons about it.

Mattrasses, also, of this striped material, stuffed and padded with coarse cotton, are still used in India as a substitute for doors and window-shutters, to keep out the heat, and are known as "purdahs." Aristobulus reported that Susiana had when he was there "an atmosphere so glowing and scorching that lizards and serpents could not cross the streets of the city at noon quickly enough to prevent their being burned to death mid-way by the heat"; that "barley spread out in the sun was roasted, as in an oven, and hopped about" (like parched peas); and that "the inhabitants laid earth to the depth of three and a half feet on the roofs of their houses to exclude the suffocating heat," so that it is not improbable that these blue and white striped "purdahs" were used in the palace of Shushan in the time of Ahasuerus.

Strabo frequently mentions this palace of Shushan, or Susa, which was in the province of Susis, or Susiana, at the head of the Persian Gulf. He tells us that when Alexander the Great became master of Persia he transferred to this residence of the Persian Monarchs everything that was precious in the land, although the palace was already almost filled with treasures and costly materials. Strabo has further been quoted as mentioning that cotton grew in Susiana and was there manufactured into cloths, but although I have searched his chapters many times I can find no such statement. It is most probable, however, that before his time cotton did grow and was manufactured in Susiana, and that it was first introduced by the Macedonians. They certainly brought into culture there before the time of Strabo another valuable plant: for we have the distinct statement of the latter that "the vine did not grow in Susiana before the Macedonians planted it both there and at Babylon."

Amidst the hurry of war and the rage for conquest Alexander always kept in view the future pacification of an invaded country; its products, therefore, were habitually ascertained and carefully noted, with a view to the increase of revenue and the development of commerce. But, beyond this, the great Macedonian conqueror, wherever he went, employed a numerous corps of scouts, and searchers, and men of science, to collect specimens of the curious animals, plants, and minerals to be found on the march. These he sent home from time to time to his great preceptor Aristotle, who was thus assisted to produce a work on Natural History which, for general accuracy of description and extent of knowledge, is a wonderful monument of scientific observation.

When by the refusal of his soldiers to proceed further than the banks of the Hyphasis (the modern Beyah), Alexander found himself obliged to yield to their wish to be led back to Persia, he determined to sail down the Indus to the ocean, and from its mouth to proceed by the Erythrean Sea to the Persian Gulf, that a communication by sea might be opened with India. His intention was that the valuable commodities of that country should thus be conveyed through the Persian Gulf to the interior parts of his Asiatic dominions, and that by the Arabian Gulf they should be carried to Alexandria (the site of which he had most judiciously selected), and thence distributed to the rest of the world.

With this object in view, he ordered a numerous fleet of boats and river-craft to be built and collected on the banks of the Hydaspes, at Bucephalia (either the modern Jhelum, or Jubalpore, some eighteen miles lower down the stream), and, when nearly two thousand vessels of various shape and size had been got together, he commenced his voyage down the Hydaspes to the Indus. The conduct of the flotilla was committed to Nearchus, an officer worthy of that important trust, though Alexander himself accompanied him in his navigation down the river. The army numbered a hundred and twenty thousand men and two hundred elephants. One third of the troops were embarked on the boats, whilst the remainder, marching in two columns, one on the right, and the other on the left side of the river, accompanied them in their progress. Retarded by various military operations on land, as well as by the slow advance of such a fleet as he conducted, Alexander did not reach the sea until more than nine months after the commencement of his journey. Having safely accomplished this arduous undertaking, he led the main body of his army back to Persia by land. The command of the fleet, with a considerable body of troops on board of it, remained with Nearchus, who, after a coasting voyage of seven months, brought it safely up the Persian Gulf into the Euphrates.

Alexander's expedition into India was no less an intelligent exploration than a successful invasion, and the western world is more indebted than is generally understood to the original genius, conspicuous foresight, political wisdom, and indefatigable exertions of that remarkable man. It was from the memoirs of his officers that Europe derived its first authentic information concerning the climate, soil, inhabitants and productions of India, and amongst the last not the least beneficial to man was cotton.

Although Scylax of Caryandra, an emissary of Darius Hydaspes, had descended the Indus to the sea about a hundred and eighty years previously (B.C. 509), other nations had derived no benefit from his investigations. But his report of the fertility, high cultivation, and opulence of the country he had passed through inflamed his master's greed, and made Darius impatient to become possessor of a territory so valuable. This he soon accomplished, and though his conquests seem not to have extended beyond the districts watered by the Indus, he levied a tribute from it which equalled in amount one-third of the whole revenue of the Persian Monarchy.

Until Alexander became master of Persia no commercial intercourse seems to have been carried on by sea between that country and India. The ancient rulers of Persia, induced by a peculiar precept of their religion which enjoined them to guard with the utmost care against the defilement of any of the "elements," and also by a fear of foreign invasion, obstructed by artificial works near their mouths the navigation of the great rivers which gave access to the interior of the country. As their subjects, however, were no less desirous than the people around them of possessing the valuable productions and elegant manufactures of India, these latter were conveyed to all parts of their dominions by land carriage. The goods destined for the northern provinces were borne on camels from the banks of the Indus to those of the Oxus, down the stream of which they were carried to the Caspian Sea, and distributed, partly by land and partly by navigable rivers, through the different countries bounded on the one hand by the Caspian, and on the other by the Euxine, or Black Sea; whilst those of India intended for the southern and interior districts were transported by land from the Caspian Gates to some of the great rivers, by which they were dispersed through every part of the country. This was the ancient mode of intercourse with India, whilst the Persian Empire was governed by its native princes; and, as Robertson says, "it has been observed in every age that when any branch of commerce has got into a certain channel, although it may not be the best or most convenient one, it requires long time and persistent efforts to give it a different direction."*

Alexander of Macedon was not a man likely to permit the existence of impediments in the way of that which he knew to be highly conducive to national progress and prosperity—namely, the expansion of commerce and facility of communication. On his return, therefore, from India to Susa, he, in person, surveyed the course of the

* Robertson's 'Historical Disquisition Concerning India.'

Euphrates and Tigris, and gave directions for the removal of the cataracts and dams, which had so long rendered the upper waters of these rivers inaccessible from the sea. His wise plans and splendid schemes were cut short by his early death, B.C. 324; but his surviving generals, though they quarrelled with each other, did their best to carry out his policy and the measures which he had concerted with so much sagacity.

His successor, Seleucus, entertained so high an opinion of the advantages to be derived from commercial intercourse with India that he organized another expedition, which must have been very successful, though no particulars of it have come down to us. He also sent to Sandracottus, King of the Prasii, an ambassador, Megasthenes, who penetrated to Palebothra (the modern Allahabad), at the confluence of the Jumna and the Ganges.

Meanwhile Ptolemy Soter, another of Alexander's generals, who had enjoyed his confidence and entered into his plans more thoroughly than any of his other officers, took possession of Egypt, and strove to secure for Alexandria the advantage of the trade with India. Some say that it was he who erected the lighthouse at the mouth of the harbour of Alexandria which was regarded as one of the seven wonders of the world, who built there the magnificent temple of Serapis, and who founded the celebrated library and museum for the benefit of learning and the cultivation of science.*

His son, Ptolemy Philadelphus, completed those works, and, further to attract the Indian trade to Alexandria, commenced to form a canal, one hundred and seventy-five feet wide, and forty-five feet deep, between Arsinoe (Suez) and the eastern branch of the Nile, by means of which the productions of India might be conveyed to Alexandria

* See Appendix H.

entirely by water. But this work was never finished, and as the navigation of the northern extremity of the Arabian Gulf (the Red Sea) was so difficult and dangerous as to be greatly dreaded, Ptolemy built a city, which he called Berenice, further down the west coast of that sea, about lat. 24°. This new city soon became the chief port of communication between Egypt and India. Goods landed there were carried by camels across the desert of Thebais to Coptos, a distance of about 320 English miles, and from there down the Nile to Alexandria, whence they were transhipped to the various countries on the Mediterranean.

Thus by the exploits and far-sighted policy of Alexander the Great were the then civilized nations of Europe made practically acquainted with calicoes, muslins, and other piece-goods—clothing materials which they had never previously seen, although probably for more than two thousand years these had been woven in the simple looms of India from the soft, white, "vegetable-lamb's wool that grew on trees"; and had during that long period supplied the principal raiment of a population of many millions.

As the Persians had an unconquerable dislike of the sea, the seat of intercourse with India was the more easily established in Egypt, and it is remarkable how soon and how regularly the commerce with the East came to be carried on by the channel in which the sagacity of Alexander had destined it to flow.

The Egyptian merchants took on board their cargoes of Indian produce at Patala (now Tatta) on the lower Delta of the Indus, at Barygaza (now Baroche, on the Nerbuddah) and in the Gulf of Cambay, and probably also at Kurrachee and Surat. As their vessels were of small burden, and as they, themselves, though sufficiently acquainted with astronomy to make some use of the stars, had no knowledge of the mariner's compass, the prudent merchantmen crept timidly along within sight of land, following the outline of every bay, and skirting the shores of Persia and Arabia and the western coast of Lower Egypt to Berenice. Though the course was tedious and the voyage prolonged, the traffic prospered, and was thus carried on for more than three centuries. When Egypt was conquered by Julius Cæsar, B.C. 30, and, after the battle of Actium, became a Roman province under Augustus, it continued undisturbed. The taste for luxury at Rome gave a new impetus to commerce with India, and at this time four hundred sailing craft were engaged in the trade.

About A.D. 50, an important discovery was made which greatly facilitated intercourse between Egypt and the East, and diminished the time occupied by the voyage. Hippalus, the commander of a vessel trading with India, noticed the periodical winds called the "monsoons," or "trade-winds," and how steadily they blew during one part of the year from the east, and during the other from the west. Having observed this to occur regularly every year, he ventured to relinquish the slow and circuitous coasting route, and stretched boldly from the mouth of the Arabian Gulf across the ocean, and was carried by the western monsoon to Musiris, on the Malabar coast. This was one of the greatest achievements in navigation in ancient history, and opened the best communication between East and West that was known for fourteen hundred years afterwards.

Arrian (who wrote A.D. 131) says that at that date Indian cottons of large width, fine cottons, muslins, plain and figured, and cotton for stuffing couches and beds, were landed at Aduli (the present Massowah), and that Barygaza was the port from which they were chiefly shipped.

The Romans also established an intercourse by land, by way of Palmyra ("Tadmor in the Wilderness"), which by means of this trade rose to great opulence; but even after the removal of the seat of government from Rome to Constantinople, in the year 329, the Roman Empire was still supplied with the productions of India by way of Egypt. The trade that might have been carried on between India and Constantinople by land was prevented by the Persians.

The Indo-Egyptian maritime traffic established by Alexander, and encouraged by Ptolemy Lagus and his son, prospered for nearly a thousand years. It survived the downfall of the Roman Empire, A.D. 476, and lasted until the conquest of Egypt by the Mahometans under Amru Benalas, the general of Caliph Omar, A.D. 634.

As no communication was carried on between Mahometans and Christians, the capture of Alexandria by the Saracens prevented the nations of Europe obtaining the products of India through Egypt, and this valuable route of international communication was abruptly stopped.

I have devoted some space to a description of the first maritime trade with India, established by the wisdom of Alexander, and suddenly arrested by Mahometan bigotry, because the history of that commerce is, more or less, the history of the cotton trade, and explains how the use of cotton and its progress westward were gradually developed and subsequently checked.

It will be convenient to make this date—the commencement of "the dark ages"—a halting-place from which to mark how far cotton and the fabrics made from it were appreciated by the nations who were chiefly benefited by the sea-carriage of Indian products in general.

The very ancient Egyptians were apparently unacquainted with cotton. At one time there was considerable discussion concerning the substance from which the swathing bandages of the mummies were woven, and some *savants* claimed to have discovered cotton amongst them. But the microscope quickly decided that question, for the character and appearance of the fibres of cotton and flax are so markedly different that any young microscopist may distinguish one from the other with ease. It was found that in every case these bandages were made of linen. Negative evidence to the same effect is furnished by the fact that no pictures or other similitude of the cotton plant has been found in Egyptian tombs, whereas accurate representations of flax occur, in its different stages of growth, harvest, and manufacture.*

The circumstance mentioned by Herodotus, that King Amasis of Egypt, in sending as a gift to Sparta a corselet padded with cotton and ornamented with gold thread, thought it a fit present from a King, and in dedicating a similar one to Minerva in her temple at Lindus considered it an offering worthy of the goddess, shows that it was at that period a novelty and a rarity. The first knowledge of cotton in Egypt may, I think, be correctly assigned to that date—about B.C. 550. Linen was the principal clothing material of the Egyptians, and the manufacture of it from flax by them is probably of as great antiquity as the growth and wearing of cotton in India. The embalmed bodies of their dead were wrapped in it during successive

* In the Grotto of El Kab are paintings representing, amongst other scenes, a field of corn and a crop of flax. Four persons are employed in pulling up the flax by the roots; another binds it into sheaves; a sixth carries it to a distance; and a seventh separates the linseed from the stem by means of a four-toothed "ripple," which he uses just in the same way as it is now used in Europe. See Hamilton's '*Ægyptiaca*,' Plate xxiii., and Yates's '*Textrinum Antiquorum*,' p. 255.

ages through a period of more than two thousand years, and their priests wore it during the same period, its clean white texture being accepted as a semblance of purity, whereas wool, taken from a sheep, was deemed a profane attire.

Flax and linen are frequently referred to in the Bible. The earliest mention of the former is in Exodus ix. 31, in the account of the plague of hail that devastated Lower Egypt B.C. 1491, and destroyed, when they were nearly ripe for harvest, the two most important crops of the Egyptians—that of the barley on which they relied for food for themselves and for export to other nations, and the flax on which they depended for their clothing and manufacturing employment. For flax was not only used for wearing apparel, but the coarser kinds were employed for making sail-cloths, ropes, nets, and for other purposes for which hemp is generally used.

It is surprising that notwithstanding the comparative proximity of Egypt to India, cotton, which had been for ages so extensively manufactured in the latter country, should have remained so long unknown or unappreciated by a people to whom it would have furnished a cheaper and more comfortable article of dress than the flax-plant. But it is certain that linen was held in favour and the use of it prevailed in Egypt till the Christian era, although the cotton fabrics imported into Berenice were gradually coming into more general wear. Pacatus mentions that Mark Antony's soldiers wore cotton in Egypt, and says that they felt so much discomfort from the heat that they could hardly tolerate light cotton clothing, even in the shade.

From a passage in Pliny's Natural History (lib. xix. cap. 1) it would appear that the cotton plant was cultivated in Upper Egypt in his day (A.D. 77), and this has been accepted as genuine and quoted by Dr. Ure* and others. But Mr. Yates, in his '*Textrinum Antiquorum*' (p. 459), shows good reason for believing that the paragraph was interpolated in the text of one of the MSS. of Pliny's work, after having been originally an annotation in the margin of an earlier copy. This explanation clears up an otherwise involved and disconnected passage, and there are other reasons besides those given by Mr. Yates for believing that his surmise is correct.

Abdollatiph, an Arabian physician who visited Egypt at the end of the twelfth century, does not mention cotton in the account which he wrote (A.D. 1203), of the plants of that country; and Prospero Alpini, the Paduan physician and botanist, who some four centuries later directed his attention to the natural history of Egypt, says † that the Egyptians then imported cotton for their use, that the herbaceous kind (*Gossypium herbaceum*), from which cotton was obtained in Syria and Cyprus, did not grow in Egypt, but that the tree kind (*G. arboreum*) was cultivated as an ornamental plant in private gardens, and in very small quantities, its down not being used for spinning.

Belon, who was in Egypt about thirty years before Alpini, makes no mention of cotton growing there; but says that he found it in Arabia, at the north of the Arabian Gulf, near Mount Sinai.

It would appear, therefore, that up to the beginning of the seventeenth century the Egyptians were importers, not cultivators, of cotton.

From a passage in the comedy 'Pausimachus' of Cecilius Statius (who died B.C. 169), quoted by Mr. Yates in the

---

* 'The Cotton Manufacture of Great Britain.'
† '*De Plantis Ægypti*,' cap. 18.

work already referred to, the Greeks seem to have been acquainted with muslins and calicoes brought from India 200 years before Christ; and about a century later the Romans adopted the Oriental custom of using cotton-cloth as a protection from the sun's rays. Ornamental coverings for tents were made from it, and awnings of striped and coloured calico were spread over the theatres, and gave welcome shade to the spectators. It was also used for sail-cloth. Cotton fabrics are frequently mentioned by the poets of the Augustan age, and by writers of a later date; but the finer qualities are almost always referred to in a manner which indicates that by the Greeks and Romans they were regarded rather as an expensive and curious production than as an article of common use. Their dress was almost entirely woollen, which, as they frequently used the bath, was always comfortable; and, for cooler wear, as Mr. Yates truly observes, "there appears no reason why cotton fabrics should have been used in preference to linen. The latter is more cleanly, more durable, and much less liable to take fire; and amongst the ancients it must have been much the cheaper of the two." In Rome and Athens the finest woven goods were extravagantly dear, for the body of the people were practically excluded from manufacturing work. This was principally carried on by slaves for the benefit of their masters, for all the great men had large establishments of slaves who understood the art of manufacturing most of the articles necessary for ordinary use. The importation of cotton and piece-goods into ancient Greece and Rome was therefore comparatively inconsiderable.

With the fall of the Roman Empire, into which Greece had previously been absorbed, art and science in Europe sank into a death-like trance which lasted for many centuries. We will therefore trace the progress of the Indian cotton trade in other directions during the long period that elapsed before science and art revived.

As India carried on a very important manufacture of cotton for home consumption, as well as for her large exports, it might be supposed that China would have been led to participate in the advantages offered by it. But, as in Egypt flax had been for many ages the raw material principally used for the clothing of the population, so in China fabrics woven from the web of the silkworm were, from the earliest times, used for the dress of all classes of the people. By authorities of high repute in China we are informed that Si-Hing, wife of the Emperor Hoang-Ti, began to breed silkworms about 2,600 years before Christ, and that the mulberry tree was cultivated to supply them with food four hundred years afterwards.

India was the country of cotton; Egypt, of flax; China, of silk; and in the two latter countries (especially in the case of the exclusive Chinese) vested interests for a long time barred the way against the adoption of the new foreign material. Cotton vestments and robes of honour were occasionally presented to the Chinese emperors by foreign ambassadors, and were highly appreciated and admired. The Emperor Ou-Ti, whose reign commenced B.C. 502, had one of these robes; but it was not till fifteen hundred years later that cotton began to be cultivated in China for manufacturing purposes. Towards the end of the seventh century the herbaceous species was grown in the gardens of Pekin, but only for the sake of its flowers. When the country was conquered by the Mongolian Tartars, A.D. 1280, the emperors of that dynasty took all possible pains to extend the culture of cotton, and imposed an annual tribute of it on several provinces. The cultivators, merchants, weavers, and wearers of silk (which included the whole nation) regarded this as a dangerous innovation seriously affecting their rights and habits, and zealously tried to maintain the established usages of the people. Eventually, however, their prejudices were overcome, and at present nine persons out of ten in China are clad in cotton raiment.

Returning to the dark ages of Europe, and the rise of the Mahometan power there, we find that by the end of the seventh century the cultivation and manufacture of cotton in Arabia and Syria had become an important industry, and had also crept along the northern coast of Africa. When, therefore, the Saracens and Moors invaded Spain and wrested it from the Goths (A.D. 712) they brought with them a knowledge of the plant and its uses. Being well skilled in agriculture, they immediately introduced in the conquered territory the cultivation of cotton, sugar, rice, and the mulberry—the latter being in favour for the use of its leaves as food for the silkworm. Looms were put to work in almost every town, and the growth and weaving of cotton were carried on with great and increasing success until the fifteenth century. Barcelona was celebrated for its cotton sail-cloth, of which it supplied a great quantity to ship-owners, and stout cotton stuffs like fustian were also qualities for which the Spanish looms were famous. Cotton paper, too, seems to have been first made by the Spanish Arabs, although about the same time it was substituted for papyrus in Egypt. A paper was likewise manufactured in Spain from linen rags which was much admired by the literary men of the time. But the religious antipathy which existed between the Moors and Christians prevented the spread of these and other Oriental arts; so that when the Moorish domination in Spain was crushed by the conquest of Grenada, in 1492, the manufactures which the Moors had introduced and fostered relapsed into barbarous neglect. The cotton plant is still found growing wild in some parts of the Peninsula. Under the influence of the Moors cotton was cultivated in Greece, Italy, Sicily and Malta, but upon their expulsion from Europe its growth was transferred to the African shores of the Mediterranean.

During the sway of the Mahometans the passage of Indian commodities to North-Western and Central Europe was so effectually barred by them that the trade dwindled, and the demand for the products of the East almost ceased. When the route through Egypt was closed, the Persians, who by that time had learned the advantages of commercial intercourse with other nations, seized the opportunity of diverting the traffic of the Persian Gulf by the Euphrates and Tigris to Bagdad, and thence across the Desert of Palmyra to the Mediterranean ports. But as Constantinople was also in the hands of the Caliphs, the roads to Europe were long and difficult. The greater part of the goods from India had, as I have mentioned (p. 58), to be carried by land on the backs of camels with the great caravans which, from time immemorial, have been the chief means of commercial intercourse between the nations of Eastern, Central, and Northern Asia, and the countries to the south and west of them.

Besides the two great caravans of pilgrims and merchants which, annually starting from Cairo and Damascus, met at Mecca, exchanged their merchandize there, and disseminated it on their return in every country they passed through, there were others consisting entirely of merchants whose sole object was commerce. These at stated seasons set out from different parts of Persia by ancient routes, on journeys of enormous length—those for the East visited India, and even the furthest extremities of China. Their average rate of travel was eighteen miles per day; and as the time of their departure and their route were both known, they were met by the people of all the countries through which they passed, for the purpose of sale, purchase, or barter. Hence the establishment, as commercial gathering-places, of the great fairs, of which that still held annually at Nijni Novgorod is a well-known example. The value of the trade thus carried on was far beyond the conception of any one who has not given especial attention to the subject. That between Russia and China, which has only been discontinued within the last few years, has been very important. In the time of Peter the Great, though the capitals of the two empires were six thousand three hundred and seventy-eight miles apart, and the route lay for more than four hundred miles through an uninhabited desert, caravans travelled regularly from one to the other. Tedious as this mode of conveyance appears, it sufficed for the traffic in Eastern produce at a period when the whole of Europe had but little time or taste for the refinements of life, and but little means of purchasing them. Nations were at that time frequently at war, the feudal barons kept their vassals under arms, a soldier's career was the only means of acquiring distinction, and luxuries obtained by commerce were looked upon as effeminate and degrading.

The arts and sciences first revived in Italy. The republics of Venice and Genoa turned their attention to commerce, and, in the year 1204, the Venetians, under Dandolo, and assisted by the soldiers of the fourth crusade, took the city of Constantinople from the Greeks, and, for a time, had the advantage of carrying on the Indian trade. They only held it, however, for fifty-seven years; for, in 1261, the Greeks, under Michael Palæologus, and aided by the Genoese, recovered possession of the city, and Genoa acquired the privileges which Venice, for a short time, had enjoyed. The Venetians then, setting aside their religious scruples, made a treaty with the Mahometans, and obtained the produce of India through Egypt.

The progress of the cotton trade, which had for so long been restricted, now became more rapid. In the fourteenth century the fustians and dimities of **Venice** and **Milan** were much esteemed, especially in Northern Europe. Half a century later the manufacture was established in Saxony and Suabia, whence it made its way into the Netherlands. At Bruges and Ghent a large trade arose, especially in the fustians which were manufactured in Prussia and Germany, and were exported thence to Flanders and Spain.

At the end of the fifteenth century two events took place within a few years of each other which formed an important epoch, not only in the history of the cotton trade, but in the history of the world—namely, the discovery of America by Columbus, and that of the passage to India round the Cape of Good Hope by Vasco da Gama. The commerce of Genoa having been supplanted by the Venetians, Christopher Columbus, a Genoese, conceived the plan of sailing to India by a new course. It having been admitted by philosophers that the world was globular, he rightly argued that any point on it might be reached by sailing westward, as well as by travelling eastward. He therefore laid his scheme, first, before the Council of the Republic of Genoa, and afterwards before the King of Portugal; but, as it was unfavourably received by both, he persuaded Ferdinand and Isabella of Spain to grant him two ships, and with these he sailed westward in search of India, on the 3rd of August, 1492. On his arrival, thirty days afterwards, at one of the Bahamas, the first land he saw after crossing the Atlantic, his vessels were surrounded by canoes filled with natives bringing cotton yarn and thread in skeins for exchange. And when he landed in Cuba, which he at first supposed to be the mainland of India, he saw the women there wearing dresses made of cotton cloth, and also found in use strong nets made of cotton cords, which the inhabitants stretched between poles and in which they slept at night. These were called "hamacas," whence comes our word "hammock." The people there had also so great a quantity of spun cotton on spindles that it was estimated there was 12,000 lbs. weight of it in a single house. Oviedo says the same of Hayti, and, at the discovery of Guadaloupe, the same year, cotton thread in skeins was found everywhere, and looms wherewith to weave it. There, as well as at Hayti and Cuba, the idols were made of cotton, and, in 1520, Fernando Magalhaens found the natives of Brazil using cotton for stuffing beds. The growth and manufacture of cotton, which were the first things brought to the notice of Columbus in the "West Indies," and which were soon afterwards found existing in various parts of South America, had apparently been handed down to those who practised them from a time far away in the past.

The Eastern Hemisphere is popularly regarded, even at the present day, as possessing a monopoly of antiquity, or, at any rate, of ancient civilization. It is not difficult to understand the mental process by which this notion is produced. In the first place the mind is hardly prepared to receive the idea that the inhabitants of countries of the existence of which we have, comparatively, so recently become aware as the continent of America should have attained to a high degree of civilization long before the natives of Britain emerged from savage barbarism. This feeling found expression in the distinctive appellations given respectively to the two hemispheres, the "Old World" and the "New World." Secondly, the only written historical records that have come down to us from the remote past relate to Europe, Asia, and Africa. But the oldest authentic history is only yesterday's news in comparison with the age of the world, and that which was called "the New World" is as old as the rest of the globe, and, apparently, was populated at quite as early a period. For in Mexico and Central America are found unmistakable proofs of the greatness and culture of former dwellers in the land. Immense piles of cyclopean masonry, of inconceivable grandeur, and incalculable antiquity; mounds and pyramids as massive as those of Egypt, huge reservoirs for water, aqueducts, ruins of public buildings, temples and palaces, tell of a powerful and wealthy nation, skilled in engineering and other sciences, and in all the important arts of civilized life. These were followed by successive races, differing from each other in habits, laws, arts, manufactures and religious worship. But all have passed away and out of memory as completely as if they had never been. We know nothing of their wars or dynasties, their prosperity or decay. Their works are their sole history. Only their ruined monuments remain to show that they once existed; and these are sometimes found in forest solitudes so far from the habitations of those who now occupy their territories, that the traveller who unexpectedly comes upon them is startled, like Crusoe by the foot-print, to find that man has been there.

In Peru, too, the companions of Pizarro found everywhere evidence of a vast antiquity, and of the former existence of a people fully equal to the Romans in grandeur of conception and skill in construction of their marvellous public works. The remains of the capital city of the Chinus of Northern Peru cover not less than a hundred and twenty square miles. Tombs, temples and palaces arise on every hand, ruined for centuries, but still traceable; immense pyramidal structures, some of them half a mile in circuit; prisons, furnaces for smelting metals, and all the structures of a busy city may still be found there. Cieça de Leon mentions having seen at Teahuanaca great buildings, and stones so large and so overgrown that it was incomprehensible how the power of man could have placed them where they were. In another place he saw enormous gateways made of masses of stone, some of which were thirty feet long, fifteen feet high, and six feet thick. The ancient Peruvians made considerable use of aqueducts, which they built with great skill of hewn stones and cement. One of these aqueducts extended four hundred and fifty miles across sierras and rivers. Their roads, macadamized with broken stone mixed with lime and asphalte, were described by Humboldt as "marvellous," and he said that none of the Roman roads he had seen in Italy, in the south of France, or in Spain, had appeared to him more imposing than the great road of the ancient Peruvians from Quito to Cuzco, and through the whole length of the empire to Chili.

These were the works of men who lived thousands of years before the times of the Incas, and amongst their manufactures was that of cotton.

In 1831, Lord Colchester brought from ancient tombs at Arica, in Peru, and placed in the British Museum, some mummy-cloths woven of cotton, the fibres of which seen under the microscope are very tortuous, and resemble those of *Gossypium hirsutum*, which is probably the primitive cotton plant of South America. The cultivation and manufacture of cotton, therefore, in the "New World" seems to have been at least coeval with the similar use of it in India.

When Pizarro conquered Peru, in 1532, he found the cotton manufacture still existent and flourishing there, for the works of the Peruvians in cotton and wool (the latter chiefly that of the vicuna) exceeded in fineness anything known in Europe at that time. He also learned that, from the foundation of the empire, at an unknown date, the dress of the Inca, or Sovereign, had always been made of cotton, and of many colours, by the "Virgins of the Sun."

When Cortez and his comrades conquered Mexico in 1519, the people had neither flax, nor silk, nor wool of sheep. They supplied the want of these with cotton, fine feathers, and the fur of hares and rabbits. The use of cotton, which had long previously existed, as is known from Aztec hieroglyphics, was as common and almost as diversified amongst the Mexicans as it is now amongst the nations of Europe. They made of it clothing of every kind, hangings, defensive armour, and other things innumerable. Cortez was so struck by the beautiful texture of some articles that were presented to him by the natives of Yucatan, that a few days after his arrival in Mexico he sent home to the Emperor Charles V., amongst other rich presents, a variety of cotton mantles, some all white, and others chequered and figured in divers colours. On the outside they had a long nap, like a shaggy cloth, but on the inside they were without any colour or nap. A number of "under-waistcoats," "handkerchiefs," "counterpanes," and "carpets" of cotton were also sent to Europe by Cortez.

Columbus's great discovery was not immediately turned to account, so far as the cotton trade was concerned, although it was destined to be most valuable to that industry at a later period. Astonishing as was his success, and great and extensive as were its results in finding a "New World" hardly inferior in magnitude to one-third of the habitable surface of the globe, he had not achieved exactly that which was the original object of his voyage—the discovery of a westerly course to India. When, therefore, only six years afterwards, a direct sea route to the East, by sailing round the Cape of Good Hope, was found, the exploit was for some time regarded as the more important of the two, because its probable effects were more easily perceptible.

The Portuguese, who had explored the west coasts of Africa which lay nearest to their own country, and had made several unsuccessful attempts to find a passage eastward, determined to make another vigorous effort to surmount the difficulty. Accordingly, on the 8th of July, 1497, a small squadron sailed from the Tagus, under the command of Vasco da Gama. After a long and dangerous voyage this navigator rounded the promontory which had for several years been the object of the hopes and dread of his countrymen, and skirting the south-east coast, arrived at Melinda, about two degrees north of Zanzibar. There he found a people so far civilized that they carried on an active commerce, not only with the nations on their own coast, but with the remote countries of Asia. Taking some of these natives on board his ships as pilots, he sailed across the Indian Ocean, and on the 22nd of May, 1498, landed at Calicut, on the Malabar coast, ten months and two days after his departure from Lisbon.

Vasco da Gama during his short stay at Melinda had little time for inquiring into the condition of the cotton trade of the country on whose shores he had landed, and it does not seem to have been forced upon his attention as it was on that of Columbus. But when Odoardo Barbosa, of

Lisbon, visited South Africa eighteen years afterwards (in 1516), he found the natives wearing clothes of cotton. In 1590, cotton cloth woven on the coast of Guinea was imported into London from the Bight of Benin, and modern travellers in the interior of Africa concur in the opinion that cotton is indigenous there, and in stating that it is spun and woven into cloth in every region of that continent. From the beauty of the dye and the designs in some of the cotton dresses, it is justly inferred to be a manufacture of very ancient standing. We have evidence, therefore, that in Africa, as well as in Asia and America, the cotton plant had a separate centre of indigenous growth, and that from a very remote period its vegetable wool was manufactured into useful and ornamental articles of clothing.*

The Portuguese took every possible precaution to secure the prize which by the courage and perseverance of their admiral they had been enabled to grasp, and to maintain the rights which priority of discovery was, in those days, supposed to confer. A chain of forts or factories was established for the protection of their trade; whilst for the extension of it they took possession of Malacca, and their ships visited every port from the Cape to Canton.

The Venetians saw with alarm the ruin that impended over them through the successful rivalry in trade of the Portuguese, but were powerless to prevent a competition against which their merchants were unable to contend. They therefore formed an alliance with the Turks under the Sultans Selim and his successor, Solyman the Magnificent, and incited them to send a fleet against the prosperous

---

* The cotton plant was also found indigenous in the Sandwich Islands, the Galapagos, etc. It is doubtful whether the cotton found in the Bornean Archipelago had not been carried eastward from India.

Portuguese. They even allowed the Turks to cut timber in the forests of Dalmatia with which to build their ships; and when twelve of these were finished, Solyman manned them with his Janissaries, and sent them to harass the Indian trade. The Portuguese met them with undaunted bravery, and, after several conflicts, vanquished the Ottoman squadron, and remained masters of the Indian Ocean.

The immediate effect of direct communication with the East by sea was the lowering of the prices of Indian produce. Commerce naturally sought the cheapest market. The trade of Venice was annihilated, and the stream of wealth that had flowed to her treasury was dried at its source. The merchandize of India was shipped from the most convenient ports, and conveyed cheaply, safely, and directly to Lisbon, and thence was distributed through Europe. A plentiful supply of Indian goods at reasonable rates caused a rapid increase in the demand for them, and amongst the trades to which this gave an impetus was that in cotton.

Up to this period no cotton was woven in England; the small quantity that was used for candle-wicks, &c., came either from Italy or the Levant. Linen was first woven in England in 1253, by Flemish hands; but for nearly a century afterwards almost all the cotton, woollen and linen fabrics consumed there were manufactured on the continent, and a great quantity of British wool was exported to Flanders and Holland. Edward III., however, gave encouragement to foreign skill, and in 1328 some Flemings settled in Manchester, and commenced the weaving of certain cloths, which, though composed of wool, were known as "Manchester cottons," and thus paved the way for the great cotton manufacture for which that part of Lancashire is now famous.

In 1560, England imported, through Antwerp, cotton brought from Italy and the Levant, as well as that carried from India to Lisbon by the Portuguese, and showed some anxiety to compete in its manufacture with foreign countries. An impulse was given to this ambition in 1585 by a fresh influx of Flemish workpeople, who, driven from their own country to escape the cruelties of the Duke of Alba during the religious persecution of the Low Countries by the Spaniards, found an asylum in England, and brought with them the skill in workmanship which adjoining States had long envied.

India, however, continued far in advance of every European country in the spinning and weaving of cotton to nearly the middle of the eighteenth century. The activity of the trade in her piece goods was looked upon as ruinous to the home manufacturer, though most profitable to the merchant, and we find Daniel Defoe, in 1708, thus lamenting, in his 'Weekly Review,' the preference for Indian chintz, calico, &c.

"It crept," he says, "into our houses, our closets, our bed-chambers; curtains, cushions, chairs, and, at last beds themselves were nothing but calicoes and Indian stuffs, and, in short, almost everything that used to be made of wool or silk, relating either to the dress of the women or the furniture of our houses, was supplied by the Indian trade. . . . The several goods brought from India are made five parts in six under our price, and, being imported and sold at an extravagant advantage, are yet capable of underselling the cheapest thing we can set about."

The Portuguese remained in undisturbed possession of the lucrative trade with India till the end of the sixteenth century, when the United Provinces of the Low Countries challenged their pretensions to an exclusive right of commerce in the East; and in 1595, the Dutch East India Company was formed. The English soon followed, and five years later (in 1600) the British East India Company was incorporated by Royal Charter. It immediately obtained from the native princes permission to establish forts and factories, and in 1624 was invested with powers of government. The Portuguese monopoly and predominance in the East was overturned and crushed, and England and Holland attained supremacy in naval power and commercial wealth.

The cotton trade did not so quickly benefit by this as might have been expected. It remained stationary for more than a century afterwards. But in 1738 commenced the history of those wonderful inventions which by giving the power of almost unlimited production to our people revolutionized the manufacturing world. England, which two centuries ago imported only £5000 worth of raw cotton, now pays more than £40,000,000 (forty million pounds) sterling every year for her supply for twelve months;* and as this supply is drawn from every quarter of the globe, she

---

* The importation of cotton into Liverpool and London in 1886 was as follows:—

	lbs.
American	1,317,562,480
Brazilian	33,832,400
Egyptian	173,340,000
West India, etc.	9,529,910
Surat	148,306,700
Madras	26,729,200
Bengal and Rangoon	32,324,600
Total	1,741,625,290

The prices of the different kinds of cotton vary according to their respective qualities, and are also influenced by the fluctuations of their market value. During 1886 the best Egyptian cotton was sometimes sold as high as 7¼d. per lb., and the inferior as low as 3⅞d. per lb.

The total value of the cotton imported during 1886 was, as I have said, rather over £40,000,000 sterling.

can appreciate the effect upon her cotton trade of the various maritime discoveries mentioned in these pages. From the country discovered by Columbus, and populated chiefly by her own offspring, England receives by far the largest portion of her requirements. The route round Cape Horn, discovered by Fernando Magalhaens in 1520, has its advantages as another road to the colonies and Eastern possessions of Great Britain. The course round the Cape of Good Hope, by which Vasco da Gama navigated his ships to Calicut, was for three and a half centuries the main road between India and Western Europe for personal intercourse, as well as the conveyance of heavy goods, such as cotton; and, though long, it was direct, and comparatively cheap. But the superiority of the first sea-route originally established by the foresight and genius of the great Macedonian conqueror was demonstrated in 1845, when Lieutenant Waghorn, a young officer in the service of the East India Company, with invincible ardour, and determined perseverance against official obstruction and innumerable obstacles, once more made Egypt the causeway between Europe and India. Alexandria, built on a site admirably chosen by its founder as a centre of commercial traffic, and placed by the prudence of his engineers just sufficiently far from the outflow of the Nile to be free from the danger of its harbour being silted up by the sediment of that muddy river, again became the port of arrival and departure : but increased skill in seamanship and the command of steam power having diminished the risk and difficulty of navigating the upper part of the Red Sea, Suez, the ancient Arsinoe, was selected for the corresponding depôt, as offering a shorter passage by land from sea to sea than the old road by Berenice, Coptos, and the Nile. Waghorn bravely carried out his scheme in the face of the most vexatious opposition and discouragement.

He built at his own expense eight halting-places in the desert between Cairo and Suez, provided carriages for passengers, and placed small steamers on the Nile and on the canal of Alexandria. At last the British and the Indian authorities, who had thrown every obstacle in his way, with an obstinate perversity which would be almost incredible if it were unique, graciously consented to countenance his plans, and to allow the mail bags to and from India to reach their destination six weeks earlier than by their former journey. Thus Thomas Waghorn brought England and her Eastern possessions by that much nearer to each other, and for this achievement deserves the gratitude of his countrymen and an honourable place in history.

The new route was, however, unsuitable to the enormous traffic in merchandize to and from the East. The unloading of cargoes at Alexandria or Suez, their "portage" across the desert, and their re-shipment on other vessels at the further side of the Isthmus, was too tedious, laborious, and expensive to be practicable; therefore the "Overland Route" was chiefly used for the rapid conveyance of the European mails, passengers, and light goods, whilst the heavy merchandize, such as cotton bales, was conveyed round the Cape as before.

In 1869, a feat of engineering was completed, the importance of which it is impossible to exaggerate. By the cutting of a deep and wide canal through the narrow strip of land which had previously barred the passage by sea round the north-eastern corner of Africa, a water-way was opened between the Mediterranean and the Red Sea, by which large ships can pass from one sea to the other without unloading their cargoes. All honour to M. de Lesseps, who, in spite of difficulties apparently insurmountable, successfully accomplished this work! He had to contend against grave political considerations, national prejudices and jealousies, religious fanaticism, vested interests, and the faithless treachery and grasping avarice of local officials. It appears to me that amidst political complications, conflicting interests, the war of tariffs, and financial arrangements, the credit and appreciation most justly due to the author of the Suez Canal have been but grudgingly given. But his posthumous fame will be lasting, and his name will be renowned in the future amongst those of the great path-finders and road-makers of the world, whose discoveries and achievements have largely benefited mankind.

The white fleeces of the wool that Alexander and his admiral saw growing on trees in India is again conveyed to Europe by the route planned for it by the great chieftain of Macedon. The water-way which he possibly suggested, and which the son of his general and confidant, Ptolemy, endeavoured, but failed, to cut, has been successfully laid open. And, although we now draw our chief supply of cotton from the western country discovered by Columbus, one result of increased facility of communication with the East, in conjunction with perfection of machinery, is that the vegetable wool coming therefrom, after giving employment to thousands of our people, and adding to our national prosperity, is returned by the same route, manufactured into various fabrics wherewith to clothe the people who cultivated it.

The subject of this chapter being the cotton trade, I need offer no apology for regarding so many of the great events of history from the point of view of their influence, especially, upon cotton as an article of commerce. Although, however, cotton is but a small item amongst the products of India, the lesson which its history forces upon all Englishmen (without distinction of religious creed, social rank, or political party) concerning the country from which it was first received in Europe and Asia is, that the possession of India confers wealth and power on her European rulers, and that Egypt is the highway to it. The nation that holds India must grasp it firmly lest it be snatched from its keeping, must guard carefully and hold strongly the road to it, and must be prepared to fight for either or both, if necessary, against any combination of enemies. For now, as in times gone by, jealous eyes are fixed upon it, and their owners only await an opportunity to put in practice that which Wordsworth makes his Rob Roy call

"the good old rule,
. . . . the simple plan,
That he shall take who has the power,
And he shall keep who can!"

APPENDIX.

A (p. 2).

Sir John Mandeville.

Sir John Mandeville, or Maundeville, was of a family that came into England with the Conqueror. He is said to have been a man of learning and substance, and had studied physic and natural philosophy. He was also a good and conscientious man, and was, moreover, the greatest traveller of his time. John Bale, in his catalogue of British writers, says of him that "he was so well given to the study of learning from his childhood that he seemed to plant a good part of his felicitie in the same; for he supposed that the honour of his birth would nothing availe him except he could render the same more honourable by his knowledge in good letters. He therefore well grounded himself in religion by reading the Scriptures, and also applied his studies to the art of physicke, a profession worthy a noble wit; but amongst other things he was ravished with a mighty desire to see the greater parts of the world, as Asia and Africa. Having provided all things necessary for his journey, he departed from his country in the yeere of Christ 1322, and, as another Ulysses, returned home after the space of thirty-four years, and was then known to a very few. In the time of his travaile he was in Scythia, the greater and lesser Armenia, Egypt, both Libyas, Arabia, Syria, Media, Mesopotamia, Persia, Chaldea, Greece, Illyrium, Tartarie and divers other kingdoms of the World, and having gotten by this means the knowledge of the languages, lest so many and great varieties and things miraculous whereof himself had been an eie-witness should perish in oblivion, he committed his whole travell of thirty-four yeeres to writing in three divers tongues—

English, French, and Latine. Being arrived again in England, having seen the wickedness of that age, he gave out this speech;—'In our time,' he said, 'it may be spoken more truly than of old that virtue is gone; the Church is under foot; the clergie is in erreur; the Devill raigneth, and Simone beareth the sway.'"

A man who in the first part of the fourteenth century could conceive, and for thirty-four years persist in carrying out, the intention of travelling from one country to another over a great part of the habitable globe, must have possessed remarkable qualifications. Indeed, his achievements were so extraordinary, and his narrative agrees in so many particulars with that of the travels of Marco Polo, that it has been suggested that he may never have gone to the East at all, but compiled his book from the journals of his predecessor. But it seems to me impossible to doubt the correctness of Mr. Halliwell's opinion that this suggestion is wholly unjustifiable, and that, after perusal of the volume, the judgment of any impartial reader would repudiate such a supposition. Sir John Mandeville met with credit and respect in his own day, and the transcriber on vellum of a small folio MS. copy of his book, written in double columns certainly not more than twenty years after his death, prefaces it in a manner which shows that he entertained no doubt concerning it.

There are several editions of Sir John Mandeville's account of his 'Voiages.' The most useful to the general reader are, 1st, that printed in London, in 1725, from a manuscript in the Cottonian collection; 2nd, a reprint of the above, with a few notes by Mr. J. O. Halliwell, and various illustrations, which are *fac-simile* copies by F. W. Fairholt, from the older editions and manuscripts in the Harleian collection, published by Lumley in 1837; and, 3rd, a reprint of this later edition, published by F. S. Ellis, in 1866.

Sir John Mandeville died at Liege on the 17th of November, 1371. His fellow-townsmen of St. Albans appear to have believed that his body was brought home to the place of his birth, and buried in St. Albans Abbey, for the following doggrel verses were inscribed as his epitaph on one of the pillars there :—

"All ye that pass by, on this pillar cast eye,
 This Epitaph read if you can ;
 'Twill tell you a Tombe once stood in this room
 Of a brave, spirited man,
 Sir John Mandevil by name, a knight of great fame,
 Born in this honoured Towne ;
 Before him was none that ever was knowne
 For travaile of so high renowne.
 As the Knights in the Temple cross-legged in Marble,
 In armour with sword and with shield,
 So was this Knight grac'd which Time hath defac'd
 That nothing but Ruines doth yield.
 His travailes being done, he shines like the Sun
 In heavenly Canaan.
 To which blessed place the Lord, of His grace,
 Bring us all, man after man."

There is no doubt, however, that Sir John Mandeville was buried in the Abbey of the Gulielmites in the town of Liege, where he died; for Abrahamus Ortelius, in his 'Itinerarium Belgiæ' (p. 16), has printed the following epitaph there set over him :—

"*Hic jacet vir nobilis Dominus Johannes de Mandeville, aliter dictus ad Barbam, Miles, Dominus de Campdi, natus de Angliâ, medicine professor, devotissimus orator, et bonorum largissimus pauperibus erogator ; qui toto quasi orbe lustrato Leodii diem viti sui clausit extremum Anno Domini* 1371, *Mensis Novembris die* 17."

Ortelius adds, that upon the same stone with the epitaph is engraven a man in armour with a forked beard, treading upon a lion, and at his head a hand of one blessing him, and these words in old French : "*Vos ki pascis sor mi, pour l'amour Deix proies por mi*"—that is, "Ye that pass over me, for the love of God pray for me." There is also a void place for an escutcheon, whereon, Ortelius was told, there was formerly a brass plate with the arms of the deceased knight engraven thereon—viz., a Lion *argent* with a Lunet *gules*, at his breast, in a Field *azure*, and a Border engraled *or*. The clergy of the Abbey also exhibited the knives, the horse-furniture, and the spurs used by Sir John Mandeville in his travels. John Weever, in his 'Ancient Funeral Monuments'

(p. 568), says that he saw the above epitaph at Liege, and also the following verses hanging near by on a tablet :—

> "*Aliud*
> *Hoc jacet in tumulo cui totus patria vivo*
> *Orbis erat: totium quem peragrasse ferunt*
> *Anglus, Equesque fuit; num ille Britannus Ulysses*
> *Dicatur, Graio clarus, Ulysse magis.*
> *Moribus, ingenio, candore, et sanguine clarus,*
> *Et vere cultor Religionis erat*
> *Nomen si quæras est Mandevil, Indus, Arabsque,*
> *Sat notum dicit finibus esse suis.*"

B (p. 8).

ODORICUS OF FRIULI.

Odoricus did not write his account of his travels with his own hand, but dictated it to his brother friar, William de Solanga, who wrote it as Odoricus related it. Having "testified and borne witness to the Rev. Father Guidolus, minister of the province of S. Anthony, in the Marquesate of Treviso (being by him required upon his obedience so to do), that all that he described he had seen with his own eyes, or heard the same reported by credible and substantial witnesses," Odoricus prepared to set out on another and a longer journey " into all the countries of the heathen." He, therefore, determined to present himself to Pope John XXII., and to obtain his benediction on his missionary enterprise. Accordingly, at the commencement of the year 1331, he left Utina with this intention. On his way, however, he was met, near Pisa, by an old man who, hailing him by his name, told him that he had known him in India, and warned him to return to his monastery, "for that in ten days thence he would depart from this present world." Having said this, he vanished from sight. Odoricus obeyed the admonition, and returned to Utina " in perfect health, feeling no crazednesse nor infirmity of body. And being in his convent the tenth day after the forsayd vision, having received the Communion, and prepared himself unto God, yea, being strong and sound of body, he happily rested in the Lord, whose sacred departure was signified to the Pope aforesaid under the hand of the public notary of Utina." Odoricus died January 14th, 1331, and was beatified.

C (p. 11).

SIGISMUND VON HERBERSTEIN.

Sigismund von Herberstein was born at Vippach, in Styria, in 1486. He distinguished himself so greatly in the war against the Turks that the Emperor entrusted him with various missions, and made him successively commandant of the Styrian cavalry, privy councillor, and president of finance of Austria. During two periods of residence at Moscow, in all about sixteen months, as ambassador from the Emperor Maximilian to the Grand Duke of Muscovy, Vasilez Ivanovich, he earnestly studied and sagaciously observed everything that came under his notice, and neglected nothing which could instruct or profit him. His work on Russia, above referred to, is universally regarded as the best ancient history of that State. He renounced public life in 1555, and died in 1556.

D (p. 14).

JULIUS CÆSAR SCALIGER.

Julius Cæsar Scaliger, born in 1484, probably at Padua, was one of the most celebrated of the many great writers of the sixteenth century. He was a man of real talent, but of unbounded vanity and unscrupulous ambition. Originally baptized "Jules," he added "Cæsar" to his name, and, to enhance his own merits by the éclat of high birth, made for himself a false genealogy, and asserted that he was the hero of adventures in which he had taken no part. In order to force himself into notice he attacked Erasmus, and in two harangues, which the latter disdained to answer, used towards him the grossest invectives. Scaliger next directed his insolent hostility against Girolamo Cardano. Jealous of the fame of the great Pavian physician and mathematician, he, in a critique containing more insults than arguments, ferociously assailed Cardano's treatise, "*De Subtilitate*"; and so exaggerated was the estimate he formed of the effect of his diatribes on the objects of his malice, that when Erasmus died, and a false rumour of the decease of Cardano was spread abroad, he believed, or affected to believe, that the death of both had been caused by his conduct towards them, and in the course of a fulsome eulogy expressed his regret for having deprived the world of letters of two such valuable lives. Scaliger died in 1558, aged seventy-five years.

E (p. 21).

JANS JANSZOON STRAUSS, OTHERWISE JEAN DE STRUYS.

Jean de Struys, in 1647, shipped at Amsterdam as sailmaker's mate on board a vessel bound to Genoa. On arriving there the ship was bought by the Republic, equipped as a privateer, and sent to the East Indies. She was, however, captured by the Dutch, and Struys took service on board a ship belonging to the Dutch East India Company, and after visiting Siam, Japan, Formosa, &c., he returned to Holland in 1681. Having stayed at home with his father for four years, he went to sea again, but finding at Venice an armed flotilla on the point of departure to fight the Turks, he joined it, was several times taken prisoner, and as often escaped or was rescued. In 1657 he returned to Holland, was married, and led a quiet life for ten years, but hearing that the Tzar was fitting out at Amsterdam some vessels to go to Persia by the Caspian Sea, "nothing," to use his own words, "could hold him back." He therefore started in a vessel bound to the Baltic, landed at Riga, and found his way overland, through Moscow and by the Oka and Volga to Astrachan. In June, 1670, the fleet in which he served set sail for the Caspian. His vessel went ashore on the coast of Daghestan, and he was made prisoner and taken to the Kan or Tchamkal of Bayance, by whom he was sold as a slave to a Persian. After passing through the possession of several masters he was bought by a Georgian, an ambassador to the King of Poland, who allowed him to purchase his freedom. On the 30th of October, 1671, he joined a caravan travelling to Ispahan, made his way to the coast, embarked for Batavia, and, after innumerable adventures, arrived in Holland in 1673, and retired to Ditmarsch, where he died in 1694. His memoirs of his life were published in Dutch, at Amsterdam, in 1677, and translated into German in the following year, and into French in 1681.

F (p. 28).

JOHN BELL OF AUTERMONY.

Furnished with letters of introduction to Dr. Areskine, chief physician and privy councillor to the Czar Peter I., Bell "embarked at London in July, 1714, on board the *Prosperity* of Ramsgate, Captain Emerson, for St. Petersburg." As the Czar was about to send an ambassador, Artemis Petronet Valewsky, to "the Sophy of Persia, Schach Hussein," Bell, by the good offices of Dr. Areskine, obtained an appointment in his suite, and set out from St. Petersburg on the 15th of July, 1715. He kept a diary, and was evidently an enlightened, discriminating and careful observer.

G (p. 52).

THE THREE BLACK CROWS.
BY DR. JOHN BYROM.

The following is the story referred to in the text. It well illustrates the process by which the first rumour concerning cotton—that "wool as white and soft as that of a lamb grew on trees"—was exaggerated to a statement that "lambs grew on certain trees," and were, therefore, partly animal and partly vegetable.

>   Two honest tradesmen, meeting in the Strand,
>   One took the other briskly by the hand.
>   "Hark ye," said he, "'tis an odd story this
>   About the crows!" "I don't know what it is,"
>   Replied his friend. "No? I'm surprised at that,—
>   Where I come from it is the common chat;
>   But you shall hear an odd affair indeed!
>   And that it happened they are all agreed:
>   Not to detain you from a thing so strange,
>   A gentleman who lives not far from 'Change,
>   This week, in short, as all the Alley knows,
>   Taking a vomit, threw up three black crows!"
>   "Impossible!" "Nay, but 'tis really true;
>   I had it from good hands, and so may you."
>   "From whose, I pray?" So, having named the man,
>   Straight to inquire his curious comrade ran.
>   "Sir, did you tell?"—relating the affair—
>   "Yes, sir, I did; and, if 'tis worth your care,
>   'Twas Mr.—such a one—who told it me;
>   But, by-the-bye, 'twas *two* black crows, *not three!*"
>   Resolved to trace so wonderous an event,
>   Quick to the third the virtuoso went.
>   "Sir,"—and so forth. "Why, yes; the thing is fact,
>   Though in regard to number not exact;
>   It was not *two* black crows, 'twas only *one!*
>   The truth of which you may depend upon;
>   The gentleman himself told me the case."
>   "Where may I find him?" "Why in—" such a place.
>   Away he went, and having found him out,
>   "Sir, be so good as to resolve a doubt;"

Then to his last informant he referred,
And begged to know if true what he had heard.
" Did you, sir, throw up a black crow ? "   " Not I ! "
" Bless me, how people propagate a lie !
Black crows have been thrown up, *three, two*, and *one;*
And here, I find, all comes at last to *none !*
Did you say nothing of a crow at all ? "
" Crow ?—crow ?—perhaps I might ; now I recall
The matter over."   " And pray, sir, what was't ? "
" Why, I was horrid sick, and at the last
I did throw up, and told my neighbours so,
Something that was—*as black*, sir, *as a crow*."

H (p. 71).

The Destruction of the Alexandrine Library.

This magnificent collection, founded by Ptolemy Soter, and added to by his successors, was twice partially dispersed before its total destruction by the Saracens. A great portion of it was burned during the siege of Alexandria by Julius Cæsar, B.C. 48. The lost volumes were in some measure replaced by Antony, who (B.C. 36) presented to Cleopatra, the library of the Kings of Pergamus. At the death of Cleopatra, Alexandria passed into the power of the Romans, and this second collection was partly destroyed by fire when the Emperor Theodosius I. suppressed paganism, A.D. 390. The Alexandrine Library met its memorable fate in 638, when, after a vigorous resistance for fourteen months, the city was taken by Amru, the general of Caliph Omar. Abdallah, the Arabian historian, and favourite of Saladin (1200), gives the following account of this catastrophe. " John Philoponus, surnamed the Grammarian, being at Alexandria when the Saracens entered the city, was admitted to familiar intercourse with Amru, and presumed to solicit a gift, inestimable in his opinion, but contemptible in that of the barbarians,—and that was the royal library. Amru was inclined to gratify his wish, but his rigid integrity scrupled to alienate the least object without the consent of the Caliph. He accordingly wrote to Omar, whose well-known answer is a notable example of ignorant fanaticism. 'If,' said he, 'these writings of the Greeks agree with the Koran they are useless, and need not be preserved; if they disagree with the book of God they are pernicious, and ought to be destroyed.' The sentence of destruction was executed with blind obedience; the volumes of paper or parchment were distributed to the 4,000 baths of the city; and so great was their number that six weeks was barely sufficient time for the consumption of this precious fuel."

INDEX.

AHASUERUS, cotton hangings in the palace of, at Shushan, 66
Alexander the Great, descent of the Indus and Hydaspes by, 68
„   „   sagacity and wise policy of, 67, 72
„   „   opens up the Euphrates and Tigris, 71
„   „   selects the site of Alexandria, 68
„   „   Europe indebted to, for the introduction of cotton, 72
Alexandria made the centre of the Indian trade, 72
„   Lighthouse, Library, and Temple of Serapis at, 71
„   destruction of the Library of—Appendix H, 105
Amasis II., Corslet padded with cotton presented to Sparta by King, 46
Aristobulus mentions "a tree bearing wool, which was carded," 47
„   report by, of the great heat at Susiana-Shushan, 66
Arrian's account of the cotton trade in his day, 73

BARNACLE Geese, the fable of, compared with that of the Borametz, 52
Borametz the, described by Sir John Mandeville, 2
„   „   „   Claude Duret, 5, 16
„   „   „   Talmudical writers, 6
„   „   „   Odoricus of Friuli, 8
„   „   „   Fortunio Liceti, 11
„   „   „   Juan Eusebio Nieremberg, 11
„   „   „   Sigismund von Herberstein, 11
„   „   „   Guillaume Postel, 13
„   „   „   Michel, the Interpreter, 13
„   „   „   Giralomo Cardano, 13
„   „   „   Julius Cæsar Scaliger, 14
„   „   „   Antonius Deusingius, 15
„   „   „   Athanasius Kircher, 21
„   „   „   Jean de Struys, 21
„   „   in verse by Guillaume de Saluste, Sieur du Bartas, 17
„   „   „   Joshua Sylvester, translator of the above, 18
„   „   „   Dr. Erasmus Darwin, 35
„   „   „   Dr. De la Croix, 36
„   „   sought for by Dr. Engelbrecht Kaempfer, 23
„   „   „   „ John Bell, of Autermony, 28, Appendix F, 103
„   „   „   „ the Abbé Chappe d'Auteroche, 30

Barometz, origin of the word, 23
  „   the fable of the, 1
  „   „   „   compared with that of the "Barnacle Geese," 52
  „   „   „   its various phases and transformations, 1, 53
Bartas, the Sieur du, lines by, on the Barometz, 17
Bell, John, seeks ineffectually the "Vegetable Lamb," 28
Borametz. *See* Barometz.
Breyn, Dr., describes to the Royal Society his Chinese artificial "Lamb," 30
British Museum, specimen of the "Scythian Lamb" in, 24, 43
Buckley, Mr., Chinese articles presented to the Royal Society by, 27
  „   „   his Chinese dog fashioned from rhizome of a fern, 27

CANAL from Suez to the East Nile commenced by Ptolemy Philadelphus, 71
  „   „   „   Aden, constructed by De Lesseps, 94
Cape route, the, discovered by Vasco da Gama, 83, 88
Cardano describes the "Vegetable Lamb," 13
  „   exposes the unreasonableness of believing the fable, 14
Central America, ancient use of cotton in, 85, 86
Chappe d'Auteroche, the Abbé seeks for the "Barometz," 30
Chinese artificial dogs made from root-stocks of ferns, 27, 28, 34, 39, 44
Columbus finds cotton in use in America, 84
Cotton, its use of great antiquity in India, 65
  „   reaches Persia from India, 66
  „   hangings of, in the palace of Ahasuerus at Shushan, 66
  „   found in use in India by Alexander the Great, 58
  „   piece-goods introduced into Europe by the Macedonians, 72
  „   shipped from Patala and Barygaza to Aduli, 72
  „   conveyed by a circuitous coasting route, 73
  „   „   in a straight course by Hippalus, 73
  „   „   by the Romans viâ Palmyra, 74
  „   the trade in, through Egypt, checked by the Saracens, 74
  „   ancient Egyptians unacquainted with, 75
  „   breast-plate padded with, sent by King Amasis to Sparta, 46, 75
  „   Mark Antony's soldiers wear, in Egypt, 76
  „   Egyptians, till the 17th century, importers, not growers of, 77
  „   in Rome and Greece manufactured by slaves, 78
  „   vestments presented to ancient Emperors of China, 79
  „   manufactured by the Moors and Saracens in Spain, 80
  „   paper made from, by the Spanish Arabs, 80
  „   manufacture in Spain relapsed after the conquest of Grenada, 80
  „   conveyed by Tartar caravans from India to Europe, 56, 57, 58, 81, 82

Cotton conveyed again through Egypt by the Venetians, 82
" manufacture in Saxony, the Netherlands, and Germany, 83
" found by Columbus in daily use in the West Indies, 84
" " Magalhaens in use in Brazil, 84
" used by the ancient Mexicans and Peruvians, 85, 86
" mummy cloths brought from ancient Peruvian tombs, 86
" imported into England in the 16th century through Antwerp, 91
" statistics, 92
" now crosses from India by the route planned by Alexander, 95
Cotton-plant, the, described by Theophrastus, 47
" " " Pomponius Mela, 48
" " " Julius Pollux, 49
" botany of the, 63
" the, indigenous to India, 64
" " noticed in India by Alexander and his army, 58
" culture of the, in China encouraged by the Mongols, 79
" " " Arabia and Syria, 77
" " " Spain by the Saracens and Moors, 80
" " " " relapsed after the conquest of Grenada, 80
" the, still grows wild in the Peninsula, 81
Cotton-wool the fleece of the " Scythian Lamb," 62
Ctesias writes of the " trees that bear wool," 46

DANIELOVITSCH, Demetrius, describes the " Vegetable Lamb " to Von Herberstein, 12
Darwin, Dr. Erasmus, lines by, on the " Barometz," 35
De la Croix, Dr., Latin lines by, on the Barometz, 36
Deusingius, Antonius, disbelieves the animal-plant monstrosity, 15
Dicksonia barometz a tree-fern, 40
" " toy dogs made from rhizomes of, by the Chinese, 41
" " does not grow in Tartary or Scythia, 44
Duret, Claude, describes the " Barometz," 3
" " avows his entire belief in the rumour, 16

EAST India Company incorporated, 92
Egypt, the route from India to Europe planned by Alexander, 68, 93, 95
" conquest of, by the Saracens, 7
" the country of flax, 75, 79
" the high road to India to be guarded, 96
Egyptian maritime traffic with the East lasted 1000 years, 74
Egyptians, the ancient, unacquainted with cotton, 75
" till the 17th century importers not growers of cotton, 77

FERNS, models of dogs made of, by the Chinese, 27, 28, 34, 39, 44
„   their economic value, 40, 41
Flemish weavers settle in Manchester, 90

GENERAL belief in the "Vegetable Lamb," 2

HEBREW, ancient, version of the fable, 6
Herberstein, Sigismund von, describes the "Vegetable Lamb," 11
Herodotus writes of trees bearing for their fruit fleeces of wool, 46
Hippalus notices the monsoons, 73

INDIA, use of cotton in, mentioned by Herodotus, 46
„   „   „   „   Ctesias, 46
„   „   „   „   Nearchus, 46
„   „   „   „   Aristobulus, 47
„   „   „   „   Strabo, 47
„   the Indo-Scythia of the ancients, 57
„   cotton indigenous to, 64
„   trade with opened by Alexander viâ Egypt, 68
„   „   viâ the Euphrates and Tigris, 71
„   „   restored to Egypt by the Venetians, 82
„   the Cape route to, discovered by Vasco da Gama, 83, 88
Indo-Scythia, identical with Scinde and the Punjaub, 57

JAPANESE artificial mermaids compared with Chinese toy-dogs, 42, 54
Jaduah, or Jeduah, the, 7

KIRCHER, Athanasius, declares the Barometz to be a plant, 21
Kaempfer, Dr. Engelbrecht, searches ineffectually for the Vegetable Lamb, 23
„   „   „   suggests that the fable refers to Astrachan lamb skins, 23

LAMB, the "Scythian," why so called, 56
„   „   „   see "Barometz."
„   „ "Tartarian," why so called, 59
„   „   „   see "Barometz."
„   „ Vegetable, its fleece cotton wool, 60
„   „   „   see "Barometz."
Lesseps, De, constructs the Suez Canal, 94
Liceti, Fortunio, says the "Vegetable Lamb" was "as white as snow," 11
Loureiro, Juan de, describes the making of artificial dogs from ferns, 44

MAGALHAENS, Fernando, discovers the route round Cape Horn, 84
Manchester, Flemish weavers settle in, 90

Mandeville, Sir John, describes the "Vegetable Lamb," 2
　　"　　"　　" biographical sketch of—Appendix A, 97
Mela, Pomponius, describes the cotton-plant, 48
Mermaids, Japanese, compared with Chinese dogs, 42, 54
Mexicans, the ancient, use of cotton by, 85 86
Michel, the Interpreter, describes the "Vegetable Lamb" and its uses, 13
Monsoons, the, noticed by Hippalus, 73
Museum, British, supposed "Scythian Lamb" in the, 24, 43
　" 　Natural History. *See* Museum, British.
　" 　Hunterian, R. Coll. Surgeons, supposed Scythian Lamb in the, 43

NEARCHUS mentions the "wool-bearing trees," 46
　　" 　descent of the Indus by, 68
Nieremberg, on the "Vegetable Lamb," 11

ODORICUS of Friuli describes the "Vegetable Lamb," 8
　　"　　　" curious incident in the life of—Appendix B, 100

PERUVIANS, the ancient, use of cotton by, 86, 87
Pliny confuses cotton with flax, 48
Pollux, Julius, describes the cotton-plant, 49
Postel, Guillaume, informs von Herberstein of the "wool-bearing plant," 13
Ptolemy Soter follows Alexander's policy and takes possession of Egypt, 71
　　"　　" founds the lighthouse, library and temple at Alexandria, 71
　　" Philadelphus commences a canal from Suez to the East Nile, 71

ROYAL Society, supposed "Scythian Lamb" laid before the, by Sir Hans Sloane, 24
Royal Society, supposed "Scythian Lamb" laid before the, by Dr. Breyn, 30

SALUSTE, Guillaume de, Sieur du Bartas. *See* "Bartas."
Scaliger, Julius Cæsar, attacks Cardano on the subject of the "Barometz," 14
Scythian Lamb, the, why so called, 56
　　"　　"　　" see "Barometz."
Scythians, the, describe snow as "feathers," 51
Scythia-Indo the same as Scinde and the Punjaub, 57
　　in Asia identical with Tartary, 57

Scythia Parva identical **with certain** districts of **Silistria and** Bessarabia, 57
Shushan, cotton hangings in the palace of Ahasuerus at, 66
Sloane, Sir Hans, lays before the Royal Society a supposed "Scythian Lamb," 24
    ,,   ,,   ,, identification of the above by, unsatisfactory, 28
    ,,   ,,   ,, bequest by, to the Nation, 43
Strabo mentions the "wool-bearing trees," 47
Strauss Jans Janszoon. *See* "Struys."
Struys, Jean de, mentions the "Barometz," 21
    ,,   ,, doubts the "animal" version of the story, 22
Suez Canal completed by De Lesseps, 94

Talmudical writers mention the "Barometz," under the name of "Jaduah," 7
Tartary identical **with** Scythia in Asia, 57
Tartar caravans, cotton conveyed by, to Europe, 56, 57, 58, 81, 82
Tartarian Lamb, the, why so called, 59
    ,,   ,,   ,, see "Barometz."
Theophrastus writes of the cultivation of the "wool-bearing tree," 47
    ,, exactly describes the cotton-plant, 48
Trees, wool-bearing, described by Herodotus, 46
    ,,   ,,   ,, Ctesias, 46
    ,,   ,,   ,, Nearchus, 46
    ,,   ,,   ,, Aristobulus, 47
    ,,   ,,   ,, Strabo, 47
    ,,   ,,   ,, Theophrastus, 47
    ,,   ,,   ,, Pomponius Mela, 48
    ,,   ,,   ,, Pliny, 48
    ,,   ,,   ,, Julius Pollux, 49

Vasco da Gama opens the Cape route to India, 83, 88
Vegetable Lamb, the, its fleece cotton wool, 60
    ,,   ,,   ,, see "Barometz."

Waghorn, Lieut., opens the route across the desert, 93
Wool-bearing trees. *See* Trees, wool-bearing.

Zavolha, the, a renowned Tartar horde, 12, 14

www.ingramcontent.com/pod-product-compliance
Lightning Source LLC
Chambersburg PA
CBHW020123170426
43199CB00009B/607